Stars and Galaxies

Stars and Galaxies
Citizens of the Universe

. . .

READINGS FROM
SCIENTIFIC AMERICAN MAGAZINE

Edited by

Donald E. Osterbrock

Lick Observatory,
University of California, Santa Cruz

W. H. FREEMAN AND COMPANY
New York

Some of the SCIENTIFIC AMERICAN articles in *Stars and Galaxies* are available as separate Offprints. For a complete list of articles now available as Offprints, write to Product Manager, Marketing Department, W. H. Freeman and Company, 41 Madison Avenue, New York, New York 10010.

Library of Congress Cataloging-in-Publication Data

Stars and galaxies : citizens of the universe : readings from Scientific American magazine / edited by Donald E. Osterbrock.
p. cm.
Includes bibliographical references.
ISBN 0-7167-2069-8 :
1. Galaxies. 2. Stars. 3. Supernovae. 4. Astronomy—History.
I. Osterbrock, Donald E. II. Scientific American.
QB857.S73 1990 89-48325
523.8—dc20 CIP

Printed in the United States of America

1 2 3 4 5 6 7 8 9 0 RRD 9 9 8 7 6 5 4 3 2 1 0

CONTENTS

Preface

In recent years SCIENTIFIC AMERICAN has published many excellent articles on stars and galaxies written by some of the outstanding creative scientists in the field. *Stars and Galaxies* is a collection of some of these articles, which, when taken together, provide a good sample of our understanding of the universe and of the methods that have enabled us to reach that understanding—an understanding that we are trying at present to go beyond.

I am very grateful to Jeremiah J. Lyons, who invited me to compile this book, and to all the editors at W. H. Freeman and Company who have worked so cooperatively with us. I am also greatly indebted to the editors of SCIENTIFIC AMERICAN for persuading such excellent scientists to write these articles and especially to the authors themselves for their authoritative and exciting chapters. Unfortunately, we could only fit twelve articles into the book, but two earlier collections, *The Universe of Galaxies,* edited by Paul W. Hodge, and *Particle Physics in the Cosmos,* edited by Richard A. Carrigan, Jr. and W. Peter Trower, contain a number of additional chapters on galaxies and cosmology, respectively.

Donald E. Osterbrock

Introduction

Astronomy, the study of the universe, is endlessly fascinating. Every tribe, nation, people and culture since the beginnings of history has tried to understand the universe and how it works. Astronomy is the science that aims at this understanding. In contrast to all the other sciences, it is explicitly concerned with the study of the entire universe since its beginning.

Study of the universe means study of the objects in the universe, that is, its inhabitants, or put another way, its citizens, since they follow the laws of nature in their motions and life processes. We could learn nothing about an empty universe; it is a concept almost impossible to define. It is only by observing the objects in the universe, recording, analyzing and interpreting the light that we receive from them, that we can begin to understand them and the nature of the universe itself.

Our sun is a star, one of the hundred billion stars that make up our galaxy. Our galaxy is a representative of the billion or so galaxies that are within reach of present-day telescopes. Ninety years ago the first estimate of this number, made by astronomer James E. Keeler on the basis of his photographic survey with a 36-inch reflecting telescope, was one hundred thousand "spiral nebulae." We have every reason to believe that larger telescopes, more sensitive detectors and better methods of discrimination against the background light of the sky at remote, dark, high-mountain observatory sites and from telescopes in space will continue to increase the number of observable galaxies.

Section I, "Pioneers," provides a background to the history of our increasing knowledge of the citizens of the universe; the other three sections of this book discuss some of the most exciting new results at the frontiers of our understanding of the universe, written by authors who made many of the discoveries they describe and who continue to follow them up.

William Herschel, whose research career is described in Chapter 1, won fame for his discovery of the previously unknown planet Uranus. More important, he made systematic observations with large telescopes he himself constructed of our galaxy of stars and nebulas and the universe of "nebulas" or galaxies. He described his observations in physical terms and tried to construct theories, some of them strikingly modern in tone, of the structure, nature and evolution of the universe. Herschel was a far-seeing pioneer and a worthy model for today's astronomers to follow in their understanding of the universe.

Another great pioneer was Henry Norris Russell, the subject of Chapter 2. He was a leader in applying the methods of physics to analyzing and interpreting the messages of starlight, the observational data collected at the telescope. A brilliant puzzle-solver, Russell exploited every new technique to understand the nature of the stars. He began with double-star orbits and ended with quantum-mechanical analyses of laboratory spectra, which he used to interpret stellar spectra in terms of the properties of their atmospheres, masses, luminosities and diameters. Astronomers today still use many of his methods, in more refined form.

Section II is concerned with galaxies, especially our own, and some of the larger structures within them. Chapter 3 deals with the very center of our galaxy, completely unobservable with optical telescopes because of the strong extinction by dust spread out along the light path between us and it, but observed in recent years in the more penetrating infrared and radio spectral regions. Here we find gas in violent motion, luminous red giant stars, molecules, dust and strong evidence for a massive, ultracompact object, probably a black hole. Our own galaxy thus has in weak form many of the properties that in strong form define an active galactic nucleus.

Globular clusters, the subject of Chapter 4, are systems of hundreds of thousands of stars within our galaxy and others. Each globular cluster is a gravitationally bound subsystem within the larger galaxy. All the stars in any individual globular cluster were born, or formed, close together in space and time. We can thus study not only the cluster itself but the aging or evolution of the stars within it and the evolution of the galaxy of which it is a part. Much of our understanding of the abundances of

the elements, how they change with time and of the life histories of stars has been derived in recent years from studies of globular clusters.

One of the exciting new developments of radio astronomy has been the discovery of cool molecular clouds, the subject of Chapter 5, which are invisible in optical radiation but detectable by the radio-frequency line radiation of their molecules. They are the dense regions of interstellar gas and dust in which new stars are born. Studies of the molecular clouds and their distribution in our galaxy and others provide important new information on star formation and on the forms and other properties of galaxies.

Chapter 6 deals with the recently discovered corona of hot gas that envelops our galaxy. It extends out from the center to immense distances in all directions. It can only be detected directly from telescopes in space, by the ultraviolet absorption lines its ions produce in the spectra of distant stars far from the galactic plane, or in even more distant galaxies and quasars, whose light shines through the corona on its way to the earth.

The inhabitants of galaxies are stars. Most of them are similar in a general way to our sun, though different in detail—most having smaller masses and lower luminosities. A few stars, however, have very high luminosities, and hence we can hope to detect them at great distances, even in other galaxies. Sections III and IV are concerned with highly luminous stars. Chapter 7 is about Epsilon Aurigae, a giant eclipsing system that certainly contains a luminous supergiant star and something that eclipses it. Astronomers' ideas of what this something is have changed over the years, but the most recent ultraviolet observations from space, combined with ground-based infrared observations, seem to show that it is a ring of dust surrounding a recently formed young star that is orbiting about the more massive supergiant.

An even more luminous object is R136a, which radiates 50 million times the sun's luminosity. It is located in the Large Magellanic Cloud and, if it is a single star, it is much more massive than any other star we know. The spectral evidence, mostly obtained in the ultraviolet range from the International Ultraviolet Explorer satellite, and the imaging data (direct photographs) and speckle interferometry results are described in Chapter 8. The conclusion is uncertain, but whether R136a is an exceptionally massive star or a rich, tightly bound cluster of massive stars, it is a fascinating citizen of the universe.

Figure I.1 STEPHAN'S QUINTET is a nearby, small and loose group of five galaxies. The photo shown here was made with the Shane 120-inch reflector at the Lick Observatory.

The most luminous stars we observe, if only briefly, are supernovas, the visible manifestation of the death throes of dying massive stars. As they explode they can become for a short time as luminous as an entire galaxy. Section IV deals with these supernovas and their remnants. Chapter 9 describes the physical processes by which an exhausted star collapses as it bounces and blows itself apart. The nuclei of the elements that form our world, including our own bodies, have been synthesized in the intense thermonuclear reactions that burn in these stars as they die and explode.

Chapter 10 tells of recent findings, many of them gained from X-ray telescopes orbiting in space, on the remnants of supernovas, intensely heated shells of expanding gas and pulsars that are the stumps of the dead stars. Studies of these remnants aid in understanding the supernova process, as well as the objects themselves.

Both Chapters 9 and 10 were written before the great supernova of 1987 was observed in the Large Magellanic Cloud. The nearest supernova by far to be seen since the application of large telescopes, modern spectrographs and advanced detectors to the study of the universe, Supernova 1987A has added immeasurably to our understanding of these celestial explosions. It is interesting to compare Chapter 11, which describes the findings from this supernova up to the summer of 1989, with Chapters 9 and 10 to see how many of the earlier ideas were confirmed, how many were extended and how few were overturned.

Chapter 12 describes the oldest pulsars in the universe, the class of dense, rapidly spinning neutron stars that did not result from supernova explosions, but came about from the accretion of matter onto either a white-dwarf star or a neutron star formed in a previous supernova. The very high-density, high-pressure, high magnetic-field physics of these objects make a fascinating final chapter to our story.

The reader should understand that he or she is on the research frontier. We are constantly learning more about the interesting citizens of the universe

described in this book. None of our knowledge is final. No science is unchanging. We build on our previous knowledge by correcting the contradictions and mistakes we find in it. Do not be surprised if you find a few such contradictions between some of the earlier chapters and the later ones. Do not be disappointed if in years to come you learn that new observational facts and new theoretical interpretations have further changed our understanding of some of the objects discussed here.

Donald E. Osterbrock

Figure I.2 CITIZENS OF THE UNIVERSE. A few of the hundred billion stars that belong to our galaxy are shown in this photo made with the Lick Observatory 20-inch astrographic telescope. They are part of the Sagittarius star cloud, which is close to the galactic plane and not too distant from the sun.

SECTION

I

PIONEERS

• • •

William Herschel and the Making of Modern Astronomy

1

He discovered thousands of stars and nebulas through telescopes that he himself built. His observations and theories expanded the bounds of astronomy to include the study of objects beyond the solar system.

• • •

Michael Hoskin
February, 1986

A typical astronomy textbook of the mid-18th century has chapters on such topics as time and celestial coordinates, and pages devoted to descriptions of the sun, moon and planets and their orbits. There is virtually nothing, however, on stellar clusters, nebulas or the large-scale structure of the universe. In contrast, textbooks published a century later deal with such subjects as a recognized part of astronomy.

The expansion in astronomical knowledge was largely brought about by the achievements of one man: William Herschel. His skill as a craftsman advanced the art of telescope construction; his dedication as an observer yielded unprecedentedly comprehensive catalogues of nebulas and stars, and his boldness as a theorist provoked the scientific study of galactic evolution. The accomplishments of this one man compelled astronomers of the 19th century to widen the compass of their field to include active study of celestial bodies outside the solar system and indeed outside our galaxy.

Yet astronomy was not Herschel's primary career. He was trained as a musician, becoming sufficiently accomplished on the oboe to join a regimental band in his native city, Hanover. In 1757, the year Herschel turned 19, the French occupied Hanover and he fled to England. He maintained himself in his new homeland first by copying music and then as a performer, conductor and composer. (He also anglicized his given name, Friedrich Wilhelm, to William.) In 1766 he was appointed organist at the fashionable Octagon Chapel in the city of Bath. It was a secure position, and although Herschel had a variety of musical duties, he was at last able to indulge his awakening intellectual interests.

Having enjoyed a monograph on the mathematical theory of harmony by Robert Smith, a professor of astronomy at the University of Cambridge, Herschel turned to Smith's previous work, a popular textbook on practical optics published in 1738. The book, *A Compleat System of Opticks,* exposed Herschel to the art of telescope making; it also contained descriptions of what could be seen in the heavens with the aid of such instruments. The practical aspect of astronomy fascinated Herschel, and he decided to try his own hand at building telescopes.

Herschel began by constructing refracting telescopes from lenses and tubes of assorted lengths, but the difficulty of handling long refractors led him to turn to reflecting telescopes, which had the added advantage that their apertures could be made wider than those of refractors. Because aperture size determines how much light a telescope can collect, faint (and distant) objects can be better examined with reflectors.

Herschel had a neighbor who made a hobby of grinding and polishing mirrors for telescopes, and he arranged to buy the neighbor's stock of tools, equipment and unfinished mirrors. Guided by Smith's book on optics, Herschel taught himself the art of finishing mirrors of speculum metal (an alloy of copper and tin) by trial and error. By the fall of 1773 he had begun to mount his own telescopic mirrors. He soon became proficient, and by January of the following year he had made a reflector with a 5½-foot focal length, a very respectable size by contemporary standards.

As the scale of Herschel's instruments increased, his ambitions grew, but the large disks he wanted to grind into concave mirrors for the telescopes he envisioned could not be cast by local foundries. Not one to give up easily, Herschel converted the basement of his house into a foundry. While he was perfecting the technique of manufacturing progressively larger mirrors, he was also making progressively better eyepieces, some of them capable of hundredfold and even thousandfold magnification. (Indeed, many of Herschel's contemporaries refused to believe his telescopes were capable of such magnifications until one of his instruments was put through a side-by-side comparison with the best instrument the Royal Greenwich Observatory had at its disposal.)

In 1776 he completed his first telescope of 20-foot focal length, which had a primary mirror 12 inches in diameter. The telescope tube was crudely slung from a pole and had to be manhandled to face in roughly the right direction. The observer peered through the eyepiece at the top of the tube while perching precariously on a ladder. Such an awkward arrangement prompted Herschel to design new telescope mountings. Two years later he had perfected a stand for his smaller telescopes that enabled him to control the motions of the telescope mechanically, by means of a system of pulleys, hinges, grooves and gears, without looking away from the eyepiece.

Herschel later incorporated many of the same mounting features into a large stand for his second 20-foot reflector, which had a mirror 18 inches in diameter (see Figure 1.1). The observer stood safely on a platform with the fine-focusing and positioning controls at hand. When necessary, the entire structure could be turned around by a single workman.

This instrument, which he called the "large" 20-foot reflector, was the model for Herschel's greatest achievement as telescope maker: a 40-foot telescope of four-foot aperture. Although it was one of the technical wonders of the 18th century, it was too cumbersome to use for frequent observation and proved to be an unsuccessful scientific instrument.

Although Herschel did sell some of his telescopes to supplement his income, the primary motivation for constructing them was a desire to personally observe objects beyond the reach of conventional astronomical instruments. In spite of his demanding job as organist and his time-consuming avocation as telescope builder, Herschel devoted countless hours to familiarizing himself with all the celestial objects his unmatched telescopes brought into view.

He assumed the role of natural historian, examining and noting the position of every specimen he came across that was brighter than a given magnitude. Like a naturalist categorizing thousands of animal or plant species, Herschel soon had to confront the problem of cataloguing the objects he had identified. Never before in the history of astronomy had anyone observed as many celestial objects as Herschel did through his powerful telescopes, let alone attempted to sort and classify them.

Herschel tackled the task with the same thoroughness and determination he had shown in building telescopes. For example, he discovered and recorded some 848 double stars (pairs of stars whose angular separation is small), with which he hoped to gauge stellar distances by means of parallax measurements. (When he reexamined some of them later, he found instances where the stars had moved around each other. This was the first direct evidence that attractive forces operate outside the solar system, as Newton had assumed but not proved.) Herschel also compiled catalogues of "the comparative brightness of stars," listing the stars in the order of decreasing apparent brightness so precisely that in future years even a slight variation in the luminosity of any star would manifest itself by throwing the sequence out of order.

Figure 1.1 "LARGE" 20-FOOT REFLECTOR, completed by Herschel in 1783, had a primary mirror 18 inches in diameter at its base. The most significant advance represented by the telescope was its unique mounting. The observer could stand on the platform at the mouth of the telescope, regardless of the telescope's declination, and have the fine-adjustment controls within arm's reach. The entire structure could be turned around by a single workman. Although the telescope was originally designed as a Newtonian reflector, with a small, flat mirror reflecting the image sideways into an eyepiece, Herschel could not accept the consequent loss of light transmission. He eliminated the small mirror and instead peered directly down the tube through an eyepiece fixed to the inside rim.

Herschel made a number of discoveries within the solar system as well. One of them did not require a telescope at all: he effectively discovered infrared rays while recording the temperature indicated by a thermometer exposed to each color of the sun's spectrum (see Figure 1.2). He noted that he in fact got the highest temperature reading just beyond the red.

Another serendipitous observation won Herschel world fame and led to his liberation from musical chores. On the evening of March 13, 1781, he was engaged in his latest and most thorough survey of the entire visible sky when he encountered an object he instantly recognized as being no ordinary star: it was not a point of light but a shining disk, whose apparent size increased in proportion to the

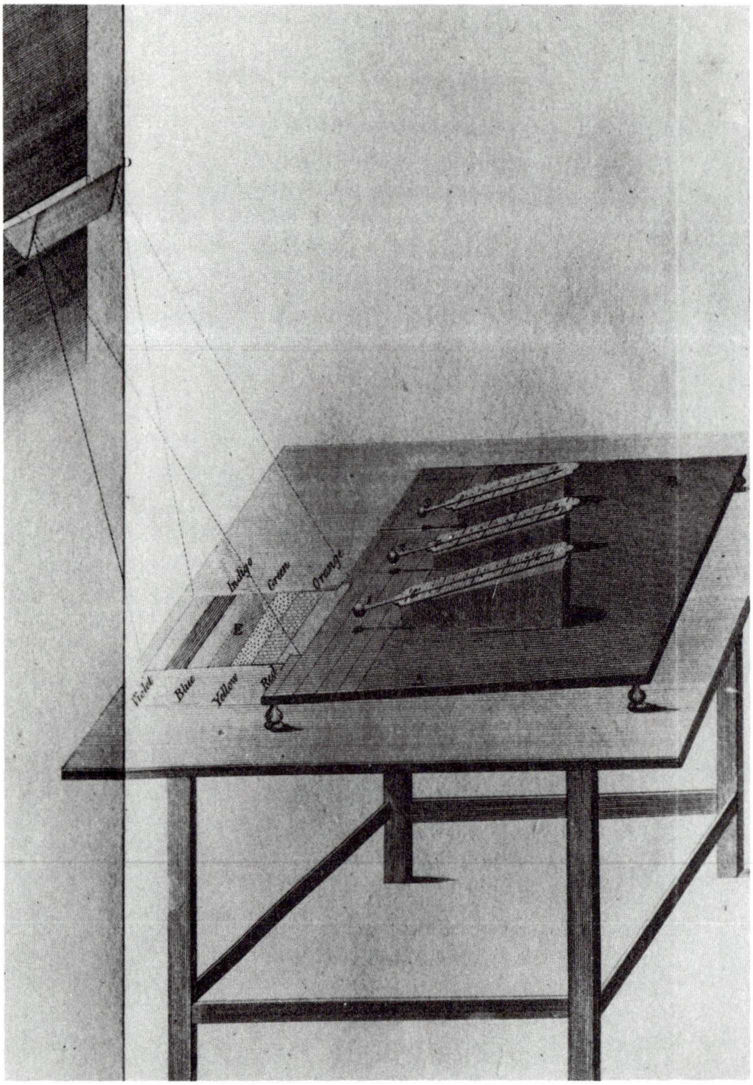

Figure 1.2 INFRARED RAYS were effectively discovered by Herschel when he noted that thermometers exposed to the various colors of the sun's light actually registered the highest temperature just beyond the red end of the visible spectrum. The experimental apparatus shown here is taken from his 1800 paper "Experiments on the refrangibility of the invisible rays of the Sun," published in *Philosophical Transactions of the Royal Society of London.*

telescopic power he applied. When he examined it again a few days later, he found it had moved; it was a member of the solar system, and presumably a comet in spite of the absence of a cometary tail. Because Herschel lacked the expertise to define the object's position accurately when he announced his discovery, professional astronomers (who were less skilled as observers and whose instruments were inferior) were exasperated by the difficulty of finding the supposed comet. When it eventually was located and its orbit was determined, it proved to be a new planet, the one we know as Uranus and the first to be discovered since antiquity.

Herschel became an international celebrity overnight and was thereupon elected a fellow of the Royal Society of London. After judicious lobbying in court circles he was awarded a royal pension by King George III, his only duties being to live near Windsor Castle and to be on call as the royal family's resident astronomer. In appreciation Herschel dubbed the planet Georgium Sidus (Georgian Star) and referred to it thereafter by that name. (In France many astronomers continued to call the planet Herschel in honor of the discoverer until the middle of the 19th century, when the name Uranus, suggested by the contemporaneous German astronomer Johann Elert Bode, finally prevailed.)

It is somewhat ironic that Herschel attained formal recognition as an astronomer for his discovery of Uranus because his major interest lay far beyond the confines of the solar system in mysterious milky patches called nebulas. (Today only regions of low-density gas and dust within galaxies are called nebulas. In Herschel's time the word was applied to any "nebulous," or fuzzy, object beyond the solar system; it included many objects now known to be galaxies.)

Herschel's lifelong fascination with nebulas is attested to by an entry on the initial page of his first observation journal: "Saw the lucid Spot in Orions Sword, thro' a 5½ foot reflector; its Shape was not

Figure 1.3 ENGRAVED PORTRAIT OF HERSCHEL depicts him holding a sketch of the planet Uranus and two of its satellites, all of which he discovered. Herschel originally named the planet Georgium Sidus (Georgian Star) in honor of the ruling British monarch, George III. Although the discovery of Uranus won Herschel worldwide fame and secured him a royal stipend, his main astronomical interest lay in star clusters and nebulas and their relation.

as Dr Smith has delineated in his Optics; tho' something resembling it; being nearly as follows." The "lucid Spot" was the Great nebula in Orion, which had been discovered and roughly sketched by the Dutch astronomer and mathematician Christiaan Huygens in 1656; Huygens' sketch had been reproduced in Smith's textbook (see Figure 1.4). The Orion nebula was observed by Herschel many times thereafter, and on one occasion he noted: "There is a visible alteration in the figure of the lucid part."

Stargazers of the early 18th century had come across a number of these mysterious objects. Preeminent among the astronomers was Charles Messier, the French comet hunter, who regarded the permanent milky patches as a source of confusion in his sweeps for comets; in 1780 he compiled a list of 68 such objects so that other astronomers would not mistake them for comets. Herschel, who found nebulas to be worthy of study in their own right, acquired Messier's list from a friend and proceeded to investigate the objects enumerated. His telescopes revealed that many of them were simply clusters of stars, but some seemed to be truly nebulous and of a different physical nature.

It was while he was searching for other such objects that Herschel made his first major discovery in stellar astronomy: "A curious Nebula, or what else to call it I do not know. It is of a shape somewhat oval, nearly circular, and with this power [a magnification of 460 times] appears to be about 10 or 15 [seconds of arc] in diameter. . . . The brightness in all the powers does not differ so much as if it were of a planetary nature, but seems to be of the starry kind." Over the course of the years Herschel found several more objects of that type. He called them planetary nebulas, a term astronomers still use today. Those mysterious glowing disks were to puzzle Herschel, and astronomers from abroad who made the pilgrimage to his home were often shown a planetary nebula and asked to give their opinion of its nature.

Herschel decided that to advance the study of nebulas further he would have to examine considerably more specimens; his large 20-foot telescope would be ideal for the purpose. He set himself to search the entire sky as visible from England, recording the position and description of as many nebulas as the reflector's 18-inch mirror could reach. For the next 20 years he would spend nights in the cold and damp by the Thames River, sweeping the sky strip by strip and shouting out positions and descriptions of nebulas to his sister Caroline, his devoted assistant throughout his career in astronomy. It was one of the most heroic campaigns in the history of observational astronomy, and it resulted in two catalogues of 1,000 nebulas each and a third one listing 500.

After Herschel's death his son extended the effort into the southern skies. The combined Herschel catalogues formed the "Catalogue of nebulae and Clusters of Stars," ultimately published in 1864; they were later enlarged by J. L. E. Dreyer into the New General Catalogue (NGC), which is still commonly referred to by astronomers today.

Herschel's work in nebular astronomy extended beyond observation and cataloguing: he theorized freely on the nature of the puzzling objects and their significance in the scheme of the universe. Indeed, the yearning to explain what he saw through his telescopes was what had driven him to observe and catalogue nebulas in the first place. As Herschel himself put it, "a knowledge of the construction of the heavens has always been the ultimate object of my observations."

In expressing his thoughts on the nature of nebulas, however, he entered into a long-standing debate among astronomers. Some held that a nebula was nothing more than a collection of stars seen as a diffuse area of brightness because of its distance from the earth, much like the Milky Way. Others believed nebulas consisted of what was referred to as nebulosity, a luminous fluid that differed from the substance of stars. Because Herschel thought he had perceived changes in the shape of the Great Nebula in Orion, he maintained that it could not be composed of stars. If a nebula were an extremely distant star system, he reasoned, even the smallest angular displacement of its components would involve movement over vast distances, and stars could not travel fast enough to account for the Orion nebula's variation in shape. On the other hand, Herschel's telescopes had shown that many of Messier's "nebulas" were star clusters. How then could one distinguish between the two appearances of nebulosity, the real and the illusory?

Herschel thought he could tell if a nebula was composed of nebulosity or of stars according to its aspect. Some nebulas had a smooth, "milky" appearance; he believed they were probably true nebulas, Others had a mottled appearance; he believed they were star clusters that could be resolved into their constituent stars by a sufficiently powerful telescope. These insights were the basis of his first

Saw the lucid Spot in Orions Sword, thro' a 5½ foot Reflector; its Shape was not as Dr. Smith has delineated in his Optics; tho' something resembling it; being nearly as follows.

From this we may infer that there are undoubtedly changes among the fixt Stars, and perhaps from a careful observation of this Spot something might be concluded concerning the Nature of it.

Figure 1.4 ORION NEBULA (M42) was depicted in Robert Smith's *A Compleat System of Opticks* (*top*) and was drawn by Herschel (*bottom*) on March 4, 1774, in his observing journal. Herschel was observant enough to note right away that "its Shape was not as D. Smith has delineated in his Optics." In the course of further examination over several years Herschel thought he detected changes in the nebula's shape. He reasoned that the Orion nebula, unlike others he observed, could not be an extremely distant cluster of stars but must instead be composed of "nebulosity," or luminous fluid.

major theoretical paper to the Royal Society, "Account of some observations tending to investigate the construction of the heavens," which he read to the society on June 17, 1784.

Just five days later Herschel trained a telescope on M17 (object number 17 in Messier's list), also known as the Omega Nebula. To his consternation he found that it seemed to contain both kinds of nebulosity, "milky" and "resolvable." "It is not of equal brightness throughout, and has one or more places, where the milky nebulosity seems to degenerate into the resolvable kind. . . . Should this be confirmed on a very fine night, it would bring on the step between these two nebulosities which is at present wanting, and would lead us to surmize that this nebula is a stupendous Stratum of immensely distant fixed stars some of whose branches are near enough to use to be visible as resolvable nebulosity, while the rest runs on to so great a distance as only to appear under the milky form."

Herschel now began to suspect that the difference between milky and resolvable nebulosity was not physical but rather was due to distance. He considered his suspicions confirmed by the configuration of M27, the Dumbbell Nebula. Stars could be made out along its axis and at the center of its two outer globular regions, but the entire object was enveloped by a diffuse aura. Herschel explained the curious structure as arising purely from distance effects: the nebula was a huge cometshaped conglomeration of stars, its nucleus pointing toward the earth and its tail flaring behind it into space. The stars that were distinguishable were those in the nucleus; the others constituted the unresolvable aura (see Figure 1.5)

In spite of the evidence to the contrary that he himself had noted in the Orion nebula, Herschel's confidence in the existence of a luminous fluid visible as milky nebulosity quickly began to wane. His new opinion was bolstered by his observation that the area of sky surrounding a nebula was generally devoid of stars. "It appeared to me remarkable that in and about the place were the many Nebulae began there was an uncommon scarcity of stars so that many fields were totally without a single star," he wrote in his observing book. "If these Nebulae should be clusters of stars it should seem as if they were collected together from the neighbouring spaces [presumably as a result of gravity's attractive force]."

Herschel theorized that all nebulas (except perhaps the planetary nebulas, which appeared to be exceptionally uniform in brightness) were composed of stars, gathered together over long periods of time by the force of gravity. To present this theory convincingly in a second paper to the Royal Society he simply suppressed all mention of the changes he thought he had detected in the Orion nebula, for these were incompatible with his new cosmogony. Instead the Orion nebula was now presented as a star system, but one so distant that it could not be resolved into stars even by Herschel's telescopes. Since it was so distant, and yet was

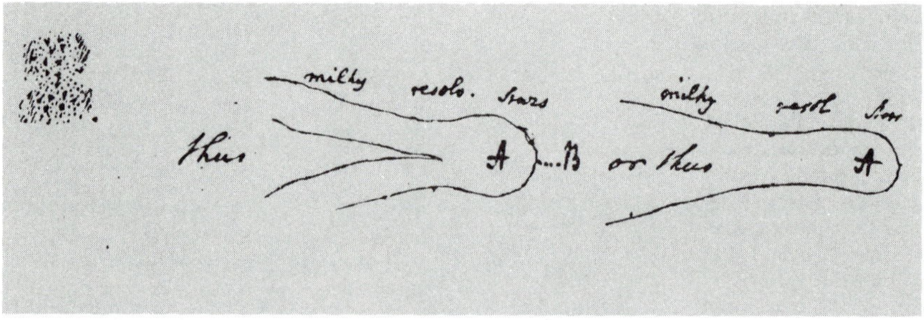

Figure 1.5 DUMBBELL NEBULA (M27) (*left*) and two alternative explanations of its appearance (*right*) were drawn by Herschel in his journal in 1784. He hypothesized that a huge, distant, comet-shaped star cluster (*A*) would look much like the Dumbbell Nebula to an observer (*B*) in front of it: the nearest stars would be seen individually, those farther away would be seen as "resolvable" nebulosity and those farthest away would be seen as "milky" nebulosity. Herschel had previously believed that the milky nebulosity was a luminous fluid quite unlike stars; M27 convinced him that this kind of nebulosity was merely the result of distance.

spread across such a wide region of sky, it must be a star system of enormous extent, one that might well "outvie our Milky Way in grandeur."

This proposition left many questions unanswered. What were the planetary nebulas, and how did they fit into the cosmogony? What was the ultimate fate of star systems condensing under gravity? Perhaps, Herschel suggested, the planetary nebulas are star systems in the final stages of gravitational collapse, and so "the stars forming these extraordinary nebulae, by some decay or waste of nature, being no longer fit for their former purposes . . . may rush at last together, and, either in succession, or by one general tremendous shock, unite into a new body." The final implosion, he maintained, could explain the new "star" recorded in 1572 by the Danish astronomer Tycho Brahe. Herschel's suggestion underscores his awareness that the force of gravity could do more than explain the stable orbits of planets and their satellites. He realized that the evolution of nebulas also had to be considered in the context of an all-pervading attractive force.

Herschel's picture of the dynamic processes that shape star clusters was incomplete, however. How were stars formed in the first place, and what happened to the matter involved in the gravitational collapse? We can be sure that Herschel was not entirely satisfied with a cosmogony in which all nebulas were simply star clusters. A solution to his difficulty manifested itself on November 13, 1790. That evening, in the course of his routine sweeps for nebulas, Herschel came on "a most singular phenomenon! A star of about the 8th magnitude, with a faint luminous atmosphere." The object was the planetary nebula NGC 1514, which has an unusually prominent central star. (Because it was not a faint disk of uniform light, Herschel in fact classified it not as a planetary nebula but as a "nebulous star.") How was he to explain it? The answer was suddenly clear to him: it was a star condensing out of a cloud of luminous fluid under the action of gravity. The revelation meant he would have to accept the existence of nebulosity and retract his theory that all nebulas are star clusters disguised by distance, but at least he could incorporate planetary nebulas into the revised cosmogony.

Herschel now believed the stellar evolutionary cycle began with a cloud of thinly scattered nebulosity that gradually breaks up into a number of smaller and denser clumps under the action of gravity. In the process of condensing, these clumps become first amorphous nebulas and then planetary nebulas. Most of the luminous substance constituting the nebulas eventually coalesces into stars, although some of it is dissipated. Such newly formed stars then congregate as a result of mutual gravitational attraction, evolving from a loose cluster of stars into a tightly packed globular cluster (see Figure 1.6). The last stage of the cycle is reached when the cluster collapses cataclysmically. The luminous material that is dispersed throughout the universe from such collapses, together with the luminosity that is constantly given off by stars and nebulas, will here and there collect into clouds of nebulosity, and so the cycle can begin anew.

As the pieces that made up the jigsaw puzzle of the universe slowly began to fall into place, Herschel even included the creation of planets in his cosmogony. He theorized that small clouds of nebulosity would sometimes be attracted toward a star and assume the form of a comet. Every time the comet passed the star some of its material would fall into the star and replenish it, while the heat of the star in turn would help to congeal the material in the comet. After many such passes the comet would be transformed into a planet. In this way Herschel enhanced the unity of his universe by having stars and planets arise jointly out of the same primordial matter.

The breadth and novelty of his theories, coupled with his naive style of writing, kept Herschel at the center of controversy during his lifetime. His contemporaries in the Royal Society argued over whether he was a genius or a charlatan. Some of the society's fellows were openly hostile, perhaps because Herschel, unlike most great observers, speculated on what he saw. Indeed, he took it as his duty to do so.

Herschel's methods and many of his theoretical investigations were ingenious. For example, he showed how statistics could be applied in astronomy when he counted the number of stars visible to the observer in various directions and, by assuming a constant density of stars, plotted a three-dimensional outline of our local star system. He was also the first to find a systematic pattern in the "proper," or individual, motions of stars; he explained this as being the result of the solar system's motion through space in the direction of the constellation Hercules.

Although many of Herschel's conclusions were legitimately criticized by his contemporaries, no one

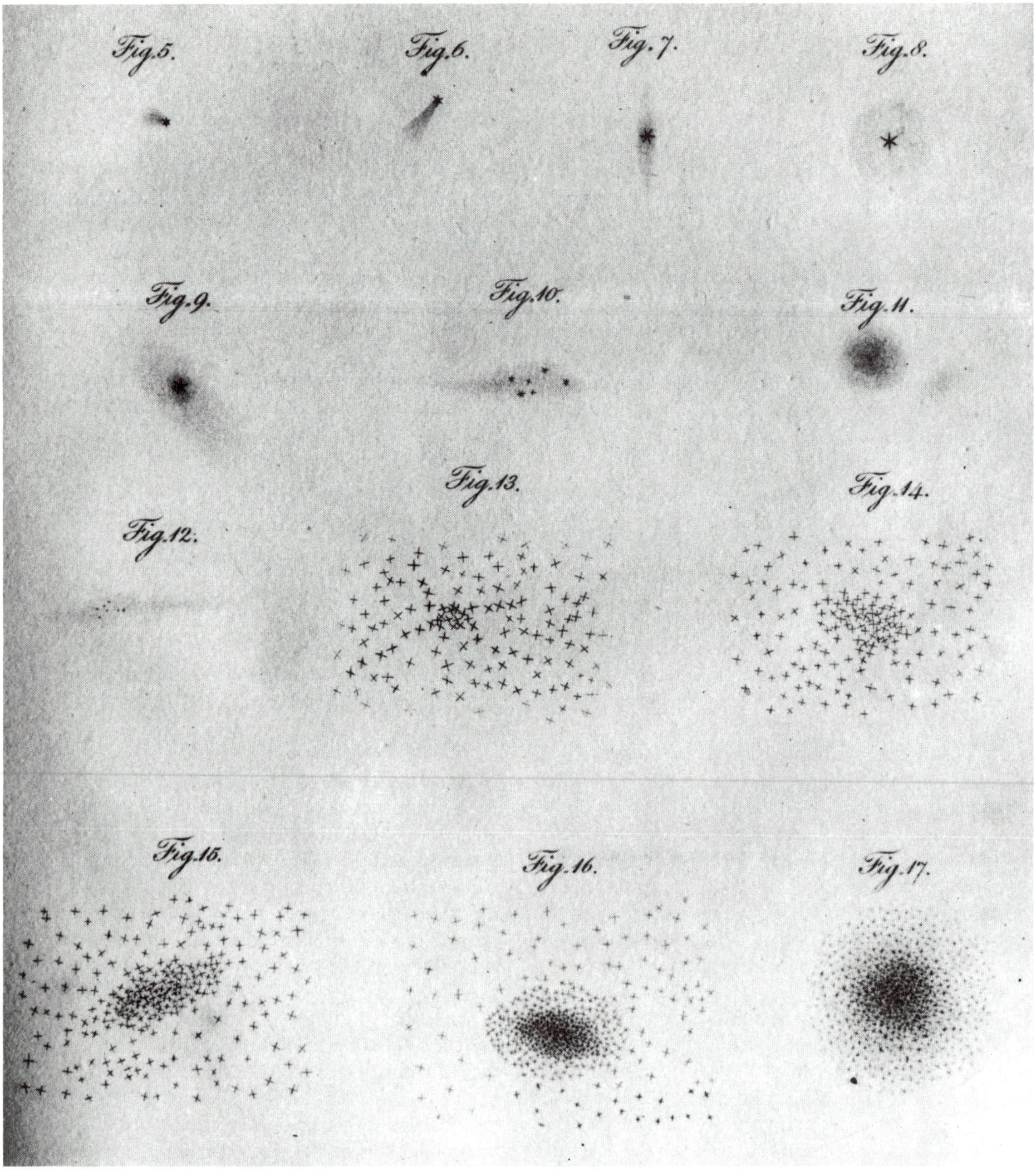

Figure 1.6 CERTAIN STARS, NEBULAS AND CLUSTERS were selected by Herschel from among the thousands he had catalogued to illustrate different stages in his theory of stellar evolution. The drawings shown here accompanied a paper by Herschel read to the Royal Society in 1814. The paper presents his theory that individual stars arise out of congealing accumulations of nebulosity and "grow" by absorbing any nebulosity that is drawn to them. The stars thus formed are then gathered into clusters by the mutually attractive force of gravity. The final stage is represented by a dense globular cluster (*bottom right*), which collapses into itself.

could deny his success in building huge telescopes, the importance of his discoveries both inside and outside the solar system and the value of his immense catalogues of double stars and nebulas. By virtue of these nontheoretical achievements he came to enjoy almost automatic right of publication in *Philosophical Transactions of the Royal Society of London.* Through this medium his theories of "the construction of the heavens" were disseminated.

His papers broached such topics as the evolution of stars and planets, the formation of galaxies and the nature of nebulas, which were all to become an accepted part of astronomy. Many of Herschel's speculations foreshadowed modern cosmological theories: they certainly displayed a perceptive appreciation of the critical role gravity plays in the workings of the universe. His most lasting legacy, however, is to be found not in particular results of his galactic and extragalactic explorations but rather in the fact the he dared to explore these regions of the cosmos in the first place. Herschel thereby ensured that the intellectual horizons of every subsequent generation of astronomers would extend far beyond the solar system that had monopolized the attention of astronomers until his time.

Henry Norris Russell

2

One of the leading astronomers of his generation, Russell understood the need to place astronomy on a firm theoretical foundation; in doing so he helped to create modern astrophysics.

• • •

David H. DeVorkin
May, 1989

In July of 1923 George Ellery Hale received the following report from one of his Mount Wilson Observatory staff about what had become a welcome summer ritual:

"Henry Norris Russell arrived, 'sailed in high,' and . . . with plenty of oil in his crankcase. The talking became a solo and continued unabated during his stay. He gave us three or four talks a week on spectral series applied to atomic. . . . Most of his time for a while before he left was devoted to working out the titanium series, according to the selections made in the furnace classification. The complexity proved greater than he expected, and he is still at the job, but the fundamental sorting out was made and it became clear what the character of the multiplets is. . . ."

This was Henry Norris Russell in his prime: a dynamo of ideas and suggestions on how to incorporate modern physics into spectroscopic astronomy. He was the Russell of the Hertzsprung-Russell diagram, of the Russell method for eclipsing binaries, of the Russell-Saunders coupling for two-electron spectra and of Russell, Dugan and Stewart's *Astronomy: Analysis of Stellar Spectra,* which helped to train two generations of astronomers. And he was the Russell who for more than 40 years contributed a monthly astronomy column to the pages of SCIENTIFIC AMERICAN.

A nervous bundle of energy, Russell was brought to Mount Wilson by Hale to exploit the store of information gathered there on laboratory and celestial spectra and to help Hale's staff inform their observational studies with the explanatory powers of modern physics. Hale hoped that Russell and other physicists would fill the gap between the laboratory bench and the observatory dome. How closing that gap became Russell's raison d'être provides insight not only into his professional life but also into a science in transition.

Figure 2.1 HENRY NORRIS RUSSELL (1877–1957) was born in Oyster Bay, N.Y., and after the age of 12 spent most of his life in Princeton. The "Dean of American Astronomers," he is remembered as the co-discoverer of the Hertzsprung-Russell diagram, which relates the brightness of stars to their colors, and of Russell-Saunders coupling, which describes two-electron interactions in atomic spectra. After 1919, when the Indian physicist Meghnad N. Saha successfully developed a theory of ionization based on quantum mechanics, Russell devoted his energy to gaining a physically rational understanding of stellar spectra.

In 1889 the 12-year-old Russell, eldest son of a minister in Oyster Bay, Long Island, was sent to live with his maternal aunt in Princeton to take advantage of the good schools there. From his mother, Eliza Norris, he had inherited a flair for mathematics and puzzle-solving and a keen sense of duty. Armed also with total recall he graduated from Princeton Preparatory School at 16 and moved on to Princeton University, where he studied mathematics and astronomy. As an undergraduate he came under the influence of Charles A. Young, a pioneer solar spectroscopist, and prepared a senior thesis on the visual classification of stellar spectra. When he went on to graduate work, his tutors included Young, the mathematician Henry B. Fine and the astronomer E. O. Lovett. This blend of mathematics and observation led to a timely doctoral thesis: a mathematical study of how Mars gravitationally perturbed the orbit of a recently discovered asteroid called Eros; the analysis would lead to a refined value for the distance between the earth and the sun.

While still a graduate student Russell established his hallmark: an acute ability to ferret out new computational techniques. He devised ways to solve for the masses of stars in visual binary systems and to determine the densities of variable stars of the Algol type, then thought by most astronomers to be a special type of eclipsing binary system. Russell soon came to realize that others were in the race too: Hendrikus J. Zwiers beat him to the visual binary technique by almost two years, and Alexander Roberts suggested the Algol technique at about the same time Russell did.

His first brushes with competition served only to push Russell harder. After completing his thesis in 1900 he broke down from overwork and recuperated during the next year by touring France. In 1902 he went on to do postdoctoral work at the University of Cambridge, where his ambitious program of theoretical and observational work on the trigonometric parallaxes of stars was again interrupted by serious illness—perhaps a nervous collapse. Leaving his work unfinished, he returned to Princeton in 1905, took a faculty post and spent the next five years attempting to find order among various observed and calculated characteristics of stars: their intrinsic brightness, colors, masses, densities, and spectra.

For this study he relied heavily on star spectra provided by Edward C. Pickering, director of the Harvard College Observatory, whose army of female assistants—in particular Annie J. Cannon—was responsible for assembling the largest collection of stellar spectra and apparent brightnesses in the world. Pickering suggested that Russell compare the absolute brightness he could derive from the parallaxes of his Cambridge stars to their spectra in the Harvard collection.

The eventual outcome of this investigation was the famous Hertzsprung-Russell diagram. (The Danish astronomer Ejnar Hertzsprung devised it independently between 1908 and 1910.) This diagram, published by Russell in 1914, shows how almost all stars can be classified according to their brightness and color and has played a crucial role in guiding theorists in their search for a theory of stellar evolution (see Figure 2.2).

During the same years Russell himself was developing a theory of stellar evolution (his giant and dwarf theory) based on an earlier hypothesis of Norman Lockyer. According to the theory a star begins its life as a vastly extended cloud of gas (a red giant), contracts and heats under self-gravitation to a critical point at which it no longer behaves like a perfect gas. It then cools (as a dense dwarf star) as it contracts further and spends the rest of its life cooling and contracting to oblivion. One must remember that in those days nuclear reactions were unknown and the theory was based simply on gravitational attraction and the kinetic theory of heat. Astronomers also had little evidence to show that stars do behave as perfect gases. The connection between modern physics and the stars was still weak.

The years 1911 and 1912 saw yet another contribution from Russell: the first quick and efficient method of deriving the orbital parameters and physical characteristics of eclipsing binary stars—later known as the Russell method for eclipsing binaries.

For his work in stellar astronomy Russell was made a full professor at Princeton in 1910, and he remained there the rest of his life. Although he shied away from politics and built no institutions, by World War I he had become a major force in American astronomy, able to dictate research agendas and direct the professional lives of workers far beyond the confines of Princeton. One early student, Harlow Shapley, became director of the Harvard College Observatory in 1921. Hale, Edwin B. Frost and Otto Struve, as successive editors of the *Astrophysical Journal*, turned to Russell for advice on what was acceptable for publication in the journal.

Although Russell's students were few, they in-

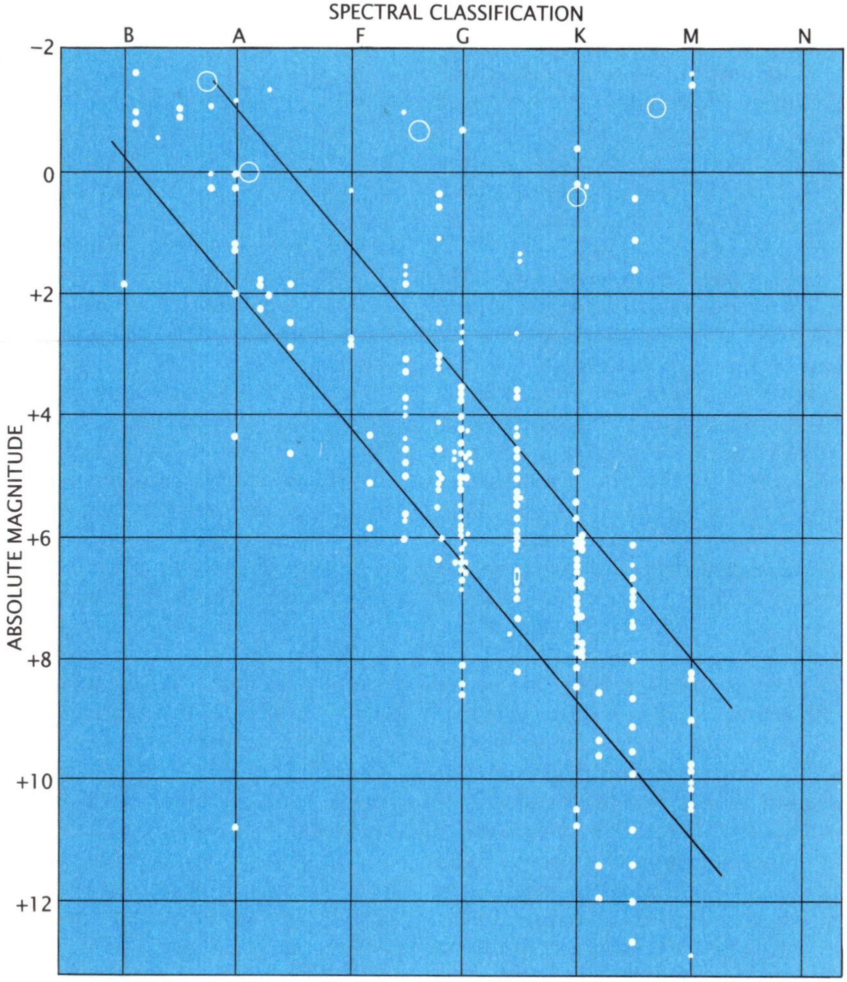

Figure 2.2 HERTZSPRUNG-RUSSELL DIAGRAM. The absolute brightness, or magnitude, of stars (*vertical axis*) is plotted against their spectral classification (*horizontal axis*), which normally ranges from blue (*O*) to red (*M*). Most stars fall along the so-called main sequence, running diagonally from the upper left to the lower right. Red giants fall in the upper right, and white dwarfs cluster at the bottom left. The diagram shown here is Russell's first published version, which appeared in *Popular Astronomy* in 1914. The circles represent various confidence levels for stellar parallaxes.

cluded the very best. Donald H. Menzel led efforts to understand the outer layers of the sun and the physics of gaseous nebulae and became director of the Harvard College Observatory in 1954. Lyman Spitzer, Jr., pioneered the application of plasma physics to the stars and became director of the Princeton University Observatory on Russell's retirement in 1947. Indeed, in surveying the American astronomical landscape of the 1930's, Struve found that the only place one could go to learn theoretical astrophysics was Princeton, at the side of Henry Norris Russell.

World War I forced Russell to turn his energies to aerial navigation and sound-ranging, but he sought diversion in planning the future of American astronomy. The country's strength lay in observational work; it boasted the world's largest telescopes and the clearest known skies. The Harvard College Observatory, Lick Observatory and Yerkes Observatory in Wisconsin were deeply committed to star surveys that had been instituted before the turn of the century.

Russell understood, however, that America lagged behind Europe in theoretical astrophysics, and he began to question the seemingly endless cataloguing projects that mapped the heavens in ever-greater detail without adequate theoretical underpinnings. Centuries earlier Newton's laws had served to describe the orbital motion of celestial bodies. What about the laws of radiation physics and quantum mechanics forged by Planck, Bohr and Einstein, which might describe the structure of stars?

Russell and others thought such questions had to be faced before more observational programs were planned. In 1916 he openly questioned the life work of his old friend and patron Pickering: "It is too often still the case that the routine work is initiated too early, before the methods are fully perfected." In 1917 Hale asked Russell to prepare the first of several research surveys for the newly established National Research Council (NRC). In his survey Russell argued that many areas in modern astronomy could be strengthened by a closer link to theoretical physics. And again, in 1917, he wrote to Pickering about the ongoing work of Pickering's assistant:

"To be quite frank it seems to me that Miss Cannon has been more concerned with what the spectra *look like* than what they *mean.* I do not think that this fact diminishes the service she has rendered to astronomy; on the contrary, her strict attention to the facts, disregarding the current theories, has given her a peculiar aptitude for her great work."

Pickering, believing that Harvard's observational program deliberations had not been adequately represented in the NRC, shot back that perhaps, in the best interests of astronomical progress, it would be good "for the dreamer to suggest to the practical man what facts he wants." The argument never erupted further. When Pickering died in 1919 Russell paid tribute to him, saying that "it was the spectroscopic information [Pickering] sent me for my parallax stars—a free gift to a young and unknown student—that started me [on the] trail that led to the whole giant and dwarf theory. I do not believe there was ever a more generous man of science."

In spite of the tribute Russell remained unshaken in his conviction that the direction of astronomy had to change. As he had written to Pickering in 1917, "It seems to me that present-day astronomy is like an army advancing with two wings, one along the line of routine observation and the other along that of investigation of principles. If the wings are not in constant touch with one another, the army will not get far." It became Russell's self-appointed task to lead American astronomy's theoretical wing and transform it into modern astrophysics.

In spite of his rhetoric, in the years that immediately followed World War I, Russell was drifting. Between 1914 and 1919 he had published 26 astronomical papers in 15 categories: binary stars, the orbit of the moon, stellar energy and evolution, stellar magnitudes and masses, parallaxes, variable stars and more. The work was good, but it certainly did not serve as the appointed signpost to the future.

Russell's 1917 NRC report shows he understood that the key to applying the revelations of Planck, Bohr and Einstein to the structure and evolution of stars lay in the analysis of stellar spectra, and yet he was stymied in his attempts to do so. At that time astronomers could identify the chemical elements producing certain absorption or emission lines only by comparing them with laboratory standards. They could classify stars according to their spectra but only in a qualitative way—by the absence or presence of lines and by the strength of the lines. The physical mechanisms behind absorption or emission features were poorly understood. Russell and others acknowledged that a difference in stellar temperatures was the primary cause for differing spectra and stellar composition probably only a secondary cause, but there were strong detractors from this view. There was as yet no physical argument for why changes in temperature created different stellar spectral features.

Astronomers were on slightly firmer ground regarding continuum spectra: the continuous, rainbowlike background on which a star's absorption or emission lines are superposed. Largely as the result of excellent colorimetry performed in Germany, there was some confidence that the amount of radiation a star emitted in a given wavelength band (its color index) bore some resemblance to Planck's famous black-body spectrum, which is the spectrum of a theoretical object whose radiative properties depend solely on temperature. But the link was as yet quite weak; moreover, many stars did not seem to behave at all like black bodies.

Russell was not the only one who recognized that a better understanding of stellar spectra was crucial. His chief counterpart in England, Arthur Stanley Eddington at Cambridge, was also thinking along the same lines. The Planck formula gives the amount of radiation that an object emits per unit area into a unit solid angle—its surface brightness. To confirm that a star behaves like a black body, one must then know the solid angle it subtends—or, equivalently, its apparent diameter. Conversely, if one assumes that a star behaves like a black body and its surface brightness is known, one can predict its apparent diameter.

In 1920 Eddington remarked in his presidential address to the British Association for the Advancement of Science that "probably the greatest need of stellar astronomy at the present day, in order to make sure that our theoretical deductions are starting out on the right lines, is some means of measuring the apparent angular diameter of stars." He then went on to estimate from its color index the temperature and surface brightness of Betelgeuse and showed that if it behaved like a black body it should subtend .051 second of arc, or 1/36,000th the angular diameter of the moon.

Confirmation came quickly. Both Eddington and Russell knew that at Mount Wilson a radical new astronomical tool was then being built under the care of Albert A. Michelson that could measure close double-star separations or even the angular diameter of stars: a giant 20-foot optical interferometer designed to sit atop the newly inaugurated 100-inch Hooker telescope. Hale, then at Pasadena, had barely put down his copy of the September 2 issue of *Nature,* where Eddington's address had been reprinted, when he took up his pen to inform Michelson that at last they had an acceptable prediction for the angular diameter of a star. By December the Michelson interferometer was put into operation and Eddington's prediction confirmed (see Figure 2.3). This was the first observational check that stars do indeed behave according to the laws of modern physics.

Russell had also made a prediction of Betelgeuse's diameter, which appeared in print in late December. It came close to the observed value, but Eddington's prediction was closer. Although Michelson later acknowledged Russell's prediction, Russell knew from this episode that the future of astronomy would largely center on the methods of the Cambridge school, which was at the cutting edge of applying physics to the study of stars.

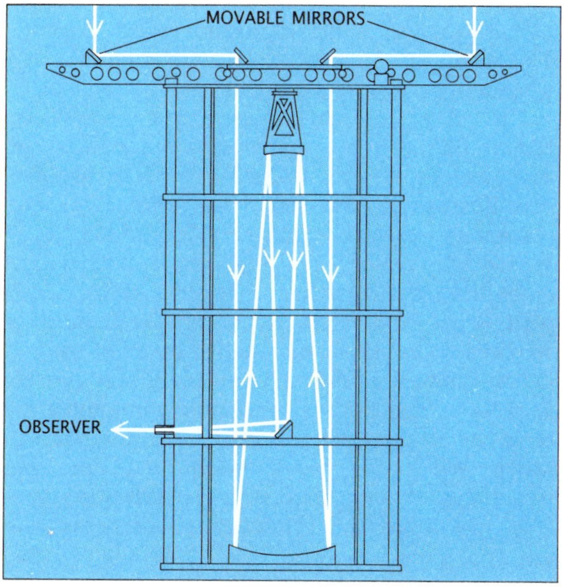

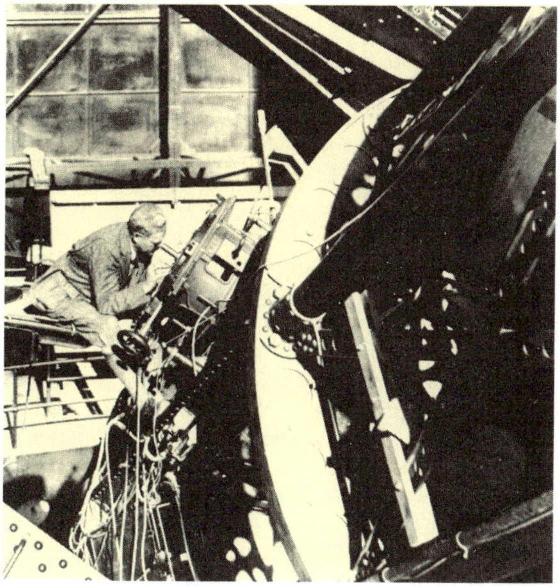

Figure 2.3 MICHELSON INTERFEROMETER was mounted on top of the 100-inch telescope on Mount Wilson in 1920. Light from a star is reflected off two movable mirrors at each end of the 20-foot-long girder; the two light beams are brought together at the focus of the telescope, where they combine, or interfere. Moving the mirrors can cause the interference fringes to vanish; the separation of the mirrors at that point is a measure of the star's angular size. With this technique the resolving power of the telescope is made equivalent to that of a telescope with a mirror 20 feet in diameter. The instrument was constructed under the supervision of Albert A. Michelson, and in late 1920 Francis G. Pease (shown here) made the first determination of the angular size of a star—Betelgeuse.

Another event of December, 1920, changed everything. That month a copy of the October issue of the *Philosophical Magazine* reached Russell. There an obscure young Indian physicist in Calcutta, Meghnad N. Saha (see Figure 2.4), boldly linked the ionization potential of a chemical element (the energy needed to detach an electron from the nucleus) and its degree of ionization to the temperature and pressure of the surrounding environment. Russell quickly recognized that this was the master key to stellar spectra he was looking for. To Walter S. Adams, one of his Mount Wilson collaborators, Russell wrote: "I believe that within a few years we may utilize knowledge of ionizing potentials, and so on, to obtain numerical determinations of stellar temperatures from spectroscopic data."

Working within the framework of the Bohr atom, Saha had applied the concepts of thermodynamic equilibrium and thermal ionization to stellar atmospheres. According to the Bohr model a photon impinging on an atom may excite an electron from one energy level to another; in the process the photon is absorbed. Conversely an electron falling from a higher energy level to a lower one emits a photon. Saha saw how these concepts could help illuminate the behavior of atoms in stellar atmospheres. In particular he showed why spectral lines of a given element are stronger at some temperatures and weaker at others.

For example, in stellar atmospheres much below 4,000 degrees Kelvin, most hydrogen atoms have their electrons in the lowest energy state. (Such stars fall into classes *K* and *M* in the familiar Harvard sequence *O, B, A, F, G, K, M,* where *O* stars are the bluest and *M* are the reddest.) Photons are unable to excite transitions between higher levels and be absorbed in the process. Therefore, absorption features corresponding to the transitions between the higher levels are absent. At about 10,000 degrees Kelvin transitions do take place between the higher levels, and absorption lines (the "Balmer series") dominate the visible spectrum. In much hotter stars (class *A*) most of the hydrogen is ionized—or exists in high states of excitation—and any transitions are above visible frequencies and hydrogen absorption features weaken again. High pressure tends to reduce the amount of ionization, and so stellar pressure also influences absorption and emission features.

Saha's October paper on stellar spectra was followed in rapid succession by two others while he was still in Calcutta. He then traveled to England, hoping to work at Cambridge, but was befriended only by Alfred Fowler of the Imperial College of Science and Technology in London. A fourth paper appeared after he had joined Fowler's laboratory. It established, as the above discussion indicates, that the *O, B, A, F, G, K, M* classification represented not only a sequence in color from blue to red but a sequence of absolute temperatures as well, *O* being the hottest and *M* being the coolest.

While Saha was in London he complained, "We have practically no laboratory data to guide us." Russell also felt the acute lack of high-quality spectroscopic and ionization data. Unlike Saha, however, he knew where to get it: from George Ellery Hale and his Mount Wilson staff.

Already in his December letter to Adams, Russell had identified ways to exploit Saha's revelations. Adams had long suspected that pressure differences in stellar atmospheres could influence the appearance of their spectra. His own method of spectroscopic parallaxes, which he developed with Anton Kohlshutter and in which certain spectral features could serve as distance indicators, hinted at the role of pressure in stellar spectra; it was now placed on a firm theoretical footing by Saha's theory. Russell pointed this out and directed Adams to his own vast store of stellar spectra at Mount Wilson, suggesting that he look for molecular hydrogen in very cool stars.

The following summer Russell himself traveled to Mount Wilson as a summer research associate to apply Saha's theory to stars. For the next two decades Russell left Princeton at least once, sometimes twice, a year for the Pasadena offices of the Mount Wilson Observatory. En route he often stopped at other observatories to advise on research, raid their plate vaults for data and lecture on a wide range of subjects. Although he maintained many of his earlier research interests, such as binary and variable stars, he added the new goal of pushing toward a theory of stellar spectra. During his first summer at Mount Wilson, he told a Berkeley audience that with Saha's theory "astronomy, physics and chemistry now had an atomic model for the emission and absorption of radiation." Russell had already used the store of spectroscopic images of the sun at Mount Wilson to show that some of Saha's predictions were correct and that overall the theory was able to explain the behavior of the elements in the solar atmosphere. He added:

"This is but a single illustration of the immense possibilities of the new field of investigation which opens up before us. A vast deal of work must be

Figure 2.4 MEGHNAD N. SAHA (1893–1956) displayed talent in mathematics and physics, but was often at odds with his superiors because of his political views and personality conflicts. His series of papers written in Calcutta and England (1919–1921) applied the concept of thermal ionization to stellar atmospheres and paved the way for an understanding of stellar spectra; they are sometimes considered the beginning of modern astrophysics. During the next two decades, he became a central figure in the creation of the National Institute of Sciences of India, the Indian Physical Society and the Indian Science News Organization.

done before it is even prospected—much less worked out, and the astronomer, the physicist, and the chemist must combine in the attack. . . . It is not too bold to hope that, within a few years, science may find itself in possession of a rational theory of stellar spectra, and, at the same time, of much additional knowledge concerning the constitution of atoms."

Russell was not the only one to see these possibilities. Saha himself pleaded for the support necessary to continue his work, and he wrote to Hale asking for the very things provided to Russell. At the same time Ralph H. Fowler and Arthur Milne in England also recognized the potential of what Saha had done and in the next few years worked to complete the theory. They noted that Saha's formula for describing the Harvard spectral classes as a sequence of absolute temperatures did not properly account for the fact that more than one element was present in a stellar atmosphere, and they rectified the omission. At the same time they refined the role of pressure.

Many others based their own work on Saha's but it was at Mount Wilson, under the coordination of Russell and Hale's staff, that the general attack on spectra took place. As a favor to Hale, Russell answered Saha's letter, outlining the planned agenda for Mount Wilson. He assured Saha that they were going to follow his lead. Saha, however, was not invited to the party.

Soon after verifying Saha's predictions Russell moved in the direction Fowler and Milne were

12 SCIENTIFIC AMERICAN **July 2, 1921**

The Heavens for July, 1921

What a Study of Atoms aud Electrons Tells Us of the Stars

By Henry Norris Russell, Ph.D.

IT is becoming more and more evident, as both sciences advance, that the astronomy of the future will be intimately associated with and dependent upon the concepts and the results of physics, and especially of that branch of physics which deals with the constitution and properties of atoms. Our knowledge within the latter field has been very greatly extended within the last decade, and many astronomical observations which before were puzzling have thereby been explained.

This is particularly true in the realm of spectroscopy. The main facts regarding the emission of light by hot bodies, and by hot gases in particular, have been known for many years; but it is only recently that we have even begun to have an idea of the processes taking place inside the atoms of the gas, which are involved.

For example, when the vapor of a given element, such as calcium or iron, is confined in a heated tube or "furnace" and observed through the end of the tube, the spectrum of the light which it emits shows certain bright lines. If the temperature is raised these lines grow stronger and new lines appear in addition. When the same metal is brought into an electric arc (which is hotter, and also subject to direct electrical action), more lines appear; while a yet more advanced stage may be reached by passing a powerful spark, fed by a source of current of high tension, between two bits of the metal; and in the spectrum from this lines may be found which were not to be observed at any of the lower stages of temperature.

Extensive studies have been made of these phenomena, and long lists of "furnace" and "spark" lines compiled, with important astronomical applications. But the physical explanation, from the atomic standpoint, lagged behind, and came only with the application of the modern quantum theory, which has been remarkably successful.

Why Are the Spectral Lines?

We have good reason to believe that an atom of any element consists of a central, and very small, nucleus, carrying a positive electrical charge, surrounded by a number of negatively charged electrons, which under the system of forces acting between them and the nucleus arrange themselves automatically in a definite pattern, probably consisting of several concentric shells or layers, at least in the heavier atoms. In the hydrogen atom there is but one electron; in helium two; in oxygen eight; in sodium eleven; in iron twenty-six; and so on up to 82 for lead and 92 for uranium. The inner electrons are held by very powerful forces, and are hard to dislodge; but a few of the outermost are relatively easy to displace, and it is these which are concerned in the chemical affinity between atoms of different sorts, and also in the production of the radiation of the visible spectrum. To pull one of these electrons away from the rest of the atom, or as it is called to ionize the atom, demands a certain expenditure of energy; and this produces an absorption of light by the gas of which this atom is a part. When some other free electron comes near to the ionized atom, it will be attracted to it (provided it does not go by too fast); and, in falling back, a corresponding amount of energy will be emitted in the form of light radiated by the gas.

Recent research has shown that this is but part of the story. There appear to be many different positions in which the electron can stop, short of being pulled clear away from the atom. The farther out it gets the more energy is required to raise it—the greatest amount of all corresponding to the complete removal of the electron, or the ionization of the atom.

Now when an electron changes from one of these states to another, light is absorbed, if it is pulled up to a "higher level" nearer the outside of the atom, or emitted if it drops to a "lower level"; and this light consists of vibrations at a perfectly definite rate, giving a sharp line in the spectrum. The most remarkable feature remains to be mentioned. The number of light vibrations per second is exactly proportional to the amount of energy which is required to pull the electron up from one position to the other, or is liberated when it comes back. The reason for this famous "quantum relation"—and indeed the reason why the various possible positions for the electron should exist at all—remains still a mystery, which is regarded by the ablest physicists as one of the hardest problems of science. But the fact has been tested in so many ways that no doubt remains.

When the spectra of the elements are studied from this standpoint it is found that the furnace lines correspond (in the case of absorption) to the raising of the electron from the very lowest "level" at which it normally is situated in the undisturbed atom to various higher levels; while the arc lines, in general, correspond to the raising of the electron from one of these higher levels to another. When light is emitted we have to do with an electron falling back over one of the same intervals.

The enhanced lines correspond to still another process. After one electron has been taken clear out of the atom, it is often possible, by a greater force, to pull a second electron out, and doubly ionize the atom. In this process too there are various possible "levels" between which the second electron may shift, and a corresponding set of lines, all quite different from the furnace or arc lines. It is even possible that an atom may lose a third or actually a fourth electron, and there is reason to suppose that some spectral lines, produced only in very violent sparks, are of this origin.

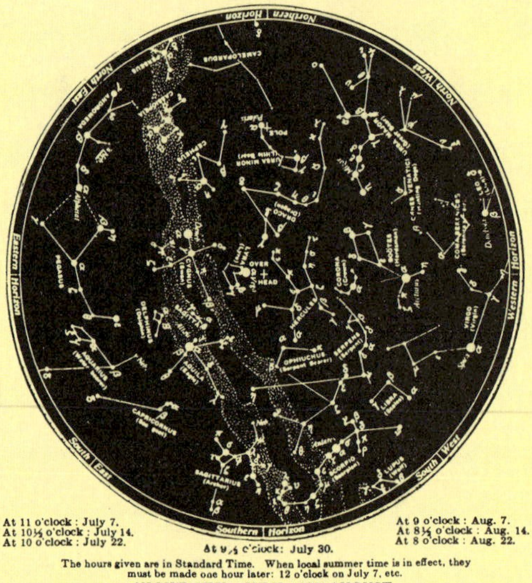

At 11 o'clock : July 7.
At 10½ o'clock : July 14.
At 10 o'clock : July 22.
At 9½ o'clock: July 30.
At 9 o'clock : Aug. 7.
At 8½ o'clock : Aug. 14.
At 8 o'clock : Aug. 22.

The hours given are in Standard Time. When local summer time is in effect, they must be made one hour later: 12 o'clock on July 7, etc.

NIGHT SKY: JULY AND AUGUST

What It Means to the Astronomer

With these ideas in mind it is very easy to see why the furnace lines are characteristic of the red stars, like Betelguese; the arc lines of yellow stars, like the sun; and the spark lines of very white stars, like Sirius. In the hot atmospheres of the stars, the atoms collide and jostle one another. The red stars are the coolest, and the collisions are the least violent, so that most of the atoms are in their undisturbed condition, and absorb only the flame lines. In the hotter atmosphere of the sun many of the atoms are jostled so that the electrons within them are raised to higher "levels" and are in a position to be raised further, with absorption of the arc lines. Finally, in the still hotter white stars most of the atoms are completely ionized, and are therefore ready to have a second electron removed, with absorption of the light corresponding to the spark lines. For some elements, such as calcium, this process occurs with relative ease; hence the spark lines of calcium—the great H and K lines in the violet—appear strongly in the sun. Helium on the other hand is the most difficult of all the elements to ionize; and the amount of energy required even to lift an electron from the lowest "level" to the next above is so great that the corresponding light vibrations are exceedingly rapid, and lie so far in the ultra-violet that all ordinarily transparent substances are opaque for them. The visible lines of helium correspond to a lifting of an electron from the second, or even a higher level to one still above, and can only be produced in an atom which has already been violently jostled, so as to throw the electron up to the second "level." This explains why the absorption lines of helium are found only in the very hot stars, like those in Orion. Spark lines of helium, corresponding to the loss of a second electron, are known; but these are found only in a very few stars which, from other evidence as well, we have reason to believe to be the hottest in the heavens.

Many beautiful applications of this theory have recently been worked out by an Indian physicist, Dr. Megh Nad Saha, of the University of Calcutta. Much of the foregoing discussion is adapted from his work, and one more instance of it may be given. The dark lines of sodium are strong in the solar spectrum. Those of potassium are present, but weak. The rare alkali metals, rubidium and caesium, show many strong lines but these do not appear in the sun at all. This has long been a puzzle, but Dr. Saha has given the solution.

Laboratory experiments have shown that it is fairly easy to remove an electron from a sodium atom, easier to get one out of a potassium atom, and still easier for rubidium and caesium. To get a second electron away from any of these atoms, after the first is gone, is however very difficult. Calculation shows that, in the sun's atmosphere, sodium vapor should be largely ionized, with however a considerable percentage remaining un-ionized atoms, which still retain one electron that may be removed by the action of light, with absorption of the well-known sodium lines. For potassium, almost all the atoms are ionized, leaving very few in a position to produce the absorption lines. Rubidium and caesium, still easier to ionize, would be completely ionized, leaving no atoms at all in a position to produce the absorption lines which are so conspicuous under the less extreme conditions of our laboratories. Hence the weakness of the potassium lines, and the absence of those of the other elements, is completely explained.

When more laboratory work has been done (largely by electrical methods) on these matters, it probably will be possible to calculate with fair precision the temperatures of the atmospheres of the various types of stars, simply from a knowledge of the degree to which the various sorts of atoms in them are ionized, as indicated by the lines in their spectra.

The Heavens

At our hour of observation Vega is almost overhead. Cygnus is high in the east, and Aquila in the southeast, a little lower. Below it lie Capricornus and Aquarius, and to the right, due south, is Sagittarius, with Scorpio to the west of it, and Ophiuchus above the latter. Bootes is the most conspicuous western constellation, with Corona above it and Hercules almost overhead. Ursa Major is in the northwest, Ursa Minor and Draco in the north, Cassiopeia and Cepheus in the northeast, and Pegasus has just risen in the east.

The Planets

Mercury is an evening star at the beginning of the

(*Continued on page 16*)

Figure 2.5 A PAGE FROM RUSSELL'S monthly column in Scientific American, explaining atomic spectra, stellar spectra and Saha's physical interpretation of them.

taking to refine the theory itself, partly to try to explain the spectroscopic anomalies unaccounted for by Saha's original version. There were two puzzles at first: the unexpected behavior of barium and the persistence of hydrogen in the spectra of all stars.

Russell found that barium was more highly ionized than sodium in the solar spectrum, which was strange because the two elements have the same ionization potential. The barium puzzle led Russell deep into the realm of physical theory to sort out the spectra of the alkaline earth elements, which include barium.

The alkaline earths distinguish themselves by having two valence, or outer, electrons instead of one. With F. A. Saunders, Russell derived a refined model for the structure of atoms in which two electrons participated in the generation of spectral lines. The rules for two-electron interaction that Russell invented during the collaboration is now called the Russell-Saunders coupling; with it the spectra of barium and the alkaline earths were explained. Even before completing this work Russell took the next step—to examine spectra from atoms with three valence electrons. He chose titanium and found yet another transition rule. Russell was elated; not only was his physics revealing how stars worked, but stellar spectra also could serve as tools to probe the mysteries of the atom.

Although barium and titanium were great successes, hydrogen remained a thorn in Russell's side. At the time astronomers believed that no one element dominated in stellar atmospheres, which were thought to be gaseous admixtures of generally heavy elements—in particular iron. Moreover, Eddington's theory of stellar structure required that the average molecular weight of the gas had to be much higher than that of hydrogen. Yet hydrogen persisted in virtually all stellar spectra.

The hydrogen puzzle led Russell and the Mount Wilson astronomers to recalibrate the solar spectral wavelengths against laboratory standards. At Princeton Russell and his indefatigable assistant Charlotte E. Moore calibrated the strengths of the solar spectral features against a new theory, developed by Russell and others, that gave the various line strengths in terms of the relative concentrations of the elements in the sun's atmosphere.

Still, the hydrogen puzzle was not resolved. In spite of the spectroscopic evidence Russell was skeptical that hydrogen dominated all stellar atmospheres. While he scratched his head over hydrogen, Russell had sent Menzel to Harvard to exploit its incomparable plate vault of stellar spectra. At the same time a young astronomer named Cecilia Payne had arrived fresh from Cambridge and Eddington, armed with insights from Fowler and Milne. She too planned to explore atomic structure with the help of the Harvard spectra and determine the elements present in stars better than Saha had been able to do. Payne succeeded in all of this in her monumental 1925 doctoral thesis, becoming the first person to recognize that hydrogen was by far the most abundant element in the atmosphere of stars.

Her conclusion was not easy to swallow. It threatened Eddington's theory of stellar structure, which prompted Russell to suggest to Payne that her result was clearly impossible. Payne dutifully followed Russell's guidance in the published form of her thesis. Privately she stuck to her conclusions.

The issue of hydrogen abundance raised its head again and again and plagued Russell's efforts to deny it through 1928, even though many of his colleagues had begun to believe that Payne's original conclusion was correct. Russell marshalled all of his forces at Princeton and Mount Wilson to make an assault on hydrogen. Finally, in a masterly 1929 paper that he referred to as his "reconnaissance of new territory," Russell gathered together all the spectroscopic evidence he and Moore had collected and declared that stellar atmospheres were, after all, mostly hydrogen.

During his final assault on hydrogen Russell was aware that Albrecht Unsöld—a student of the theoretical physicist Arnold Sommerfeld and completely conversant with the most modern forms of quantum theory—was able to derive absolute-abundance information from spectroscopic line profiles, something Russell had never tried to do. Unsöld was also confirming that stellar atmospheres were composed chiefly of hydrogen, and Russell knew that the young German's techniques were far more powerful than his own.

Unsöld's mastery of the new quantum physics represented the future of stellar-atmosphere studies. He and a host of other Europeans would ultimately refine Russell's first crude estimate of the relative abundances in the solar atmosphere. Russell was delighted with these extensions of his work, but it was apparent that the new "quantum mechanics" were rapidly taking over. Even in the mid-1920's, as Russell worked to find a theory of multiplet spectra

(spectra produced by closely clustered atomic energy levels), Sommerfeld, H. Hönel and R. de L. Kronig were hot on the trail—and in fact they beat him into print.

Such theoretical races made Russell aware of the army of European physicists who were then attacking atomic structure. Russell knew he was outnumbered and to a great extent outdistanced. In part this stemmed from an aversion to the direction in which physics was heading. Russell never felt comfortable with the complex mathematical formalism of quantum mechanics and was always happier with what he called the astronomical model of the Bohr atom, right down to its metaphor of "spin." Much later Russell still referred to Heisenberg's uncertainty principle as a "Principle of Limited Measurability," following Max Born. Neither was he ever comfortable with the wave-particle duality of matter, although he was willing to apply either model to "practical problems."

After the mid-1920's Russell preferred to let others handle developments in theory. He continued to admire the power and generality of the Bohr model and its ability to provide rules for the calculation of atomic spectra. Along with so many other spectroscopic physicists of his generation, Russell contented himself with puzzle-solving: applying the Bohr model to spectra in order to unravel the structure of atoms.

Russell's role as a pioneer in quantitative astrophysics was that of one who pointed the way. He was a transition figure who never made the transition fully himself; his students did, however, and much of Russell's lasting influence came from them, astronomers such as Spitzer and Menzel and those they trained in turn.

Russell's many roles prompted Shapley to knight him with the title "Dean of American Astronomers." In Russell was found an unusual blend of two classic scientific styles: the hedgehog and the fox—the deep versus the broad. He was sympathetic to the need for programmatic observation but uncomfortable with vast projects uninformed by theory. Russell often saw through the fog to suggest fruitful lines of research that others might carry out. As Cecilia Payne-Gaposchkin once said late in life, "Henry Norris Russell knew a good thing when he saw it." By taking advantage of these "good things" and also conveying to others the need for long-term systematic enquiry informed by physical theory, Russell accrued wide influence. In a survey of astronomers made in 1946, Russell was listed most frequently as an especially stimulating teacher; citations of his work still averaged about 50 per year in the 1960's and 1970's, long after his death at 79 in 1957.

Editor's Note The author acknowledges permission from the Princeton University Library, manuscript division, to quote correspondence from the Henry Norris Russell papers.

SECTION

II

GALAXIES

. . .

The Central Parsec of the Galaxy

3

Infrared and radio observations indicate that the center of our galaxy harbors an ultracompact object, possibly a massive black hole, embedded in a dense, swirling mass of stars, gas and dust.

• • •

Thomas R. Geballe
July, 1979

In the early 1930's Karl Jansky, a young physicist employed by the Bell Telephone Laboratories in New Jersey, built the first radio telescope to assist him in identifying sources of interference that could hamper transatlantic radio-telephone communication. One major source of static Jansky discovered was a "steady hiss," as he described it, that came from the sky and was strongest in a direction far to the south of his instrument. Over the course of time he observed that the source of this interference followed the motion of the fixed stars and hence originated outside the solar system. The direction of peak intensity, toward the constellation Sagittarius, coincided well with what was already considered by astronomers to be the location of the center of our galaxy.

Jansky's observations were not only the first radio-astronomical observations but also the first detection at any wavelength of the nucleus of the galaxy, which is shrouded by clouds of interstellar dust. Later, with the benefit of the advances in electronics during World War II, radio astronomy earned recognition as a full-fledged scientific enterprise. In the 1960's and 1970's infrared technology, which had lagged behind its radio counterpart, began to provide sensitive detectors and spectrometers that could be used in conjunction with large optical telescopes. Because of their ability to "penetrate" the intervening clouds of dust, radio and infrared astronomy have contributed most of the knowledge that has been gained about the central region of the galaxy. In the past few years continued advances in both fields have enabled astronomers to begin to explore in detail the innermost and very densest region.

Galactic nuclei are dense in mass solely as a consequence of gravitation. The galaxies in the universe are thought to have formed from density nonuniformities in the gas ejected by the "big bang" in which the universe was created. In time the nonuniformities contracted gravitationally to form galaxies, and within them smaller clouds collapsed by gravitation to form stars and other objects. All the objects in a galaxy are also subject to a net gravitational pull toward its nucleus. Although the rotation of the galaxy slows the infall of matter, and momentary outbursts (for example supernovas) may eject matter from the center, frictional forces inevitably cause the center to acquire a mass density much higher than that of the outer regions of the galaxy. The ultraluminous objects known as quasars, which can be observed at vast distances, may be an extreme result of the process. The center of even a normal-looking spiral galaxy such as our own, however, might be expected to harbor almost the entire galactic astronomical zoo: main-sequence nebulas, dust clouds, neutron stars and perhaps even a black hole —all packed together in a tight swarm.

Galactic nuclei are the probable source of the enormous expulsions of material that have been observed in some distant galaxies. It appears that the energy necessary for such expulsions can come only from dense galactic nuclei. Our galaxy also shows evidence of past explosive events, albeit relatively mild ones, in the large clouds of gas observed by radio astronomers to be moving outward from the nucleus.

Perhaps the most exciting discoveries currently being made about the nucleus of our galaxy concern its central parsec, a volume only 3.26 light-years in diameter. At the distance of the center one parsec subtends an angle of 20 seconds of arc, which is equal to the angle subtended on the earth by a medium-size crater on the moon. Within that volume, which would fit comfortably between the sun and its nearest stellar neighbor, there are millions of stars and a variety of other objects whose nature astronomers are just beginning to understand. There is also evidence that an amount of material a few million times more massive than the sun is concentrated at the very heart of the central parsec, perhaps in the form of a black hole. Although the black hole cannot be observed directly, its existence is suggested by the behavior of the surrounding material.

I have chosen the unit of a parsec for this account not because it has any inherent virtue over some other unit, such as the light-year, but simply because the parsec is the astronomer's habitual unit of distance. One parsec is the distance from the earth to an object whose position as observed from the earth seems to shift back and forth from its average position by an angle of one arc-second (1/36,000 degree) as a result of the parallax as the earth travels around the sun. The distance to the center of the galaxy is about 10,000 parsecs, or about 20 times greater than the distance to the well-known Great Nebula in Orion. On the other hand, the center of the galaxy is about 70 times closer to the solar system than the nucleus of any other galaxy is. Therefore one should be able to perceive the details of the nucleus of our galaxy 70 times more clearly than those of the nucleus of any other galaxy. One problem, of course, is that when a telescope is pointed toward the center of our galaxy, it does not distinguish among objects lying in the foreground, in the nucleus or in the background. There are various stratagems, however, for making such distinctions and for deciding which features are actually at the galactic center.

The largest telescopes and the most sensitive photographic emulsions are able to record only the merest trace of visible light emitted from the millions of stars packed within the nucleus of our galaxy (see Figure 3.1). Because of intervening clouds of dust only one photon of visible light in about 10^{11} photons survives the journey of 30,000 years from the galactic center to the earth. Even the rare surviving photons are swamped by the exposure-limiting background radiation consisting of visible photons scattered in the earth's atmosphere. In the infrared region of the spectrum, however, a much larger fraction of the photons complete their long journey. In the radio region photons from the galactic nucleus are only rarely scattered and absorbed. At the other end of the spectrum, in the X-ray region, photons also travel unimpeded, but they do not survive passage through the earth's atmosphere and therefore can be recorded only by instruments carried to high altitude. Our atmosphere, however, is transparent at most radio and some infrared wavelengths. As a result most of what has been learned of the central region of the galaxy has been provided by radio telescopes and by sensitive infrared detectors placed at the foci of reflecting telescopes.

If our galaxy were free of dust, one could use a large optical telescope to record the entire central parsec of the galaxy on a single photographic plate with a resolution of about a second of arc. With a telescope in an artificial satellite the resolution could be improved by a factor of 10 or 20. That improvement might be just sufficient to resolve some 100,000 stellar objects within a diameter of 20 arc-seconds, corresponding to the central parsec of the galaxy. If the central parsec actually contains a few million stellar objects, as now seems probable, they would merge into bright clumps, and perhaps into a single brilliant mass, even if they were photographed from an artificial satellite at high resolution.

One can obtain an impression of what a dust-free view of the galactic nucleus might look like from

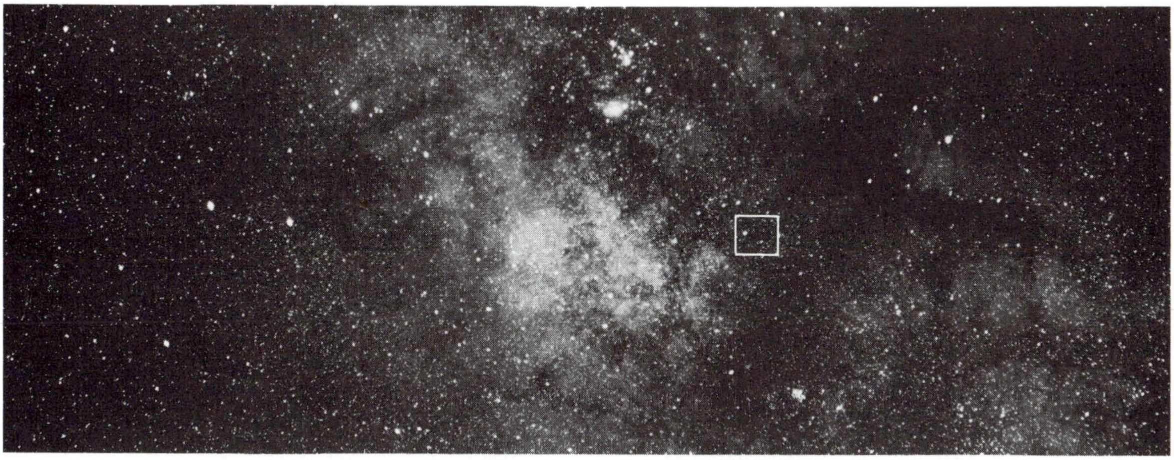

Figure 3.1 CENTER OF THE GALAXY lies hidden in the southern Milky Way in the direction of the constellation Sagittarius. The distance from the solar system to the center of the galaxy is about 10,000 parsecs, or slightly more than 30,000 light-years. The light from the millions of stars in the nucleus of the galaxy is completely obscured by interstellar dust. Infrared and radio emission from the galactic nucleus, however, is detectable. The rectangle, which is about twice the diameter of the full moon, outlines area that is shown in the infrared map of Figure 3.3.

photographs taken of the nucleus of the Great Nebula in Andromeda, the nearest spiral galaxy similar to our own (see Figure 3.2). Because of the way the two galaxies are oriented in space the nucleus of the Andromeda galaxy can be photographed along a line of sight that is largely free of dust, which is concentrated in the central planes of the two systems.

The fact that the center of our galaxy is invisible at visible wavelengths has undoubtedly been a stimulus to efforts to observe it at other wavelengths. In order to obtain a "picture" at infrared and radio wavelengths it is necessary to scan successive strips of the sky (by moving the telescope, for example) while recording point by point the number of photons striking the detector. In this way one can build up a television-like image with a resolution that is simply the area of the sky the detector sees at any instant. The limit to the resolution is established by the effective aperture of the telescope: the higher the ratio of aperture to wavelength, the higher the maximum achievable resolution. In most cases data acquired in this way are converted into contour maps on which the lines connect points of equal brightness in the same way that lines on a topographic map connect points of equal elevation. Recently with computer assistance it has become fairly easy to convert such data into color-coded two-dimensional pictures.

Before proceeding to the central parsec of the galaxy I shall describe briefly what has been learned from radio and infrared studies of the extended nuclear region in which it lies. Radio contour maps of the region have been made at many wavelengths. All the maps reveal a distribution of radiation that is elongated along the galactic equator. The position of the strongest peak, known to radio astronomers as Sagittarius *A*, is also common to all the maps, and it has long been suspected of coinciding with the galactic center (see Figure 3.3). Sagittarius *A* must have contributed to the steady hiss detected by Jansky, although his instrument lacked the resolution to distinguish it from other peaks nearby.

The sources of emission that appear on radio maps of the galactic center are the clouds of hot interstellar gas known as H II regions. These regions, which are common in our galaxy, consist of ionized atoms, that is, atoms that have been stripped of one or more of their electrons. The atoms, mostly those of hydrogen, are ionized by the ultraviolet radiation from nearby stars (or other sources) that are much hotter than the sun. The positively charged atomic fragments and the negatively charged free electrons in an H II region emit radio waves over a broad

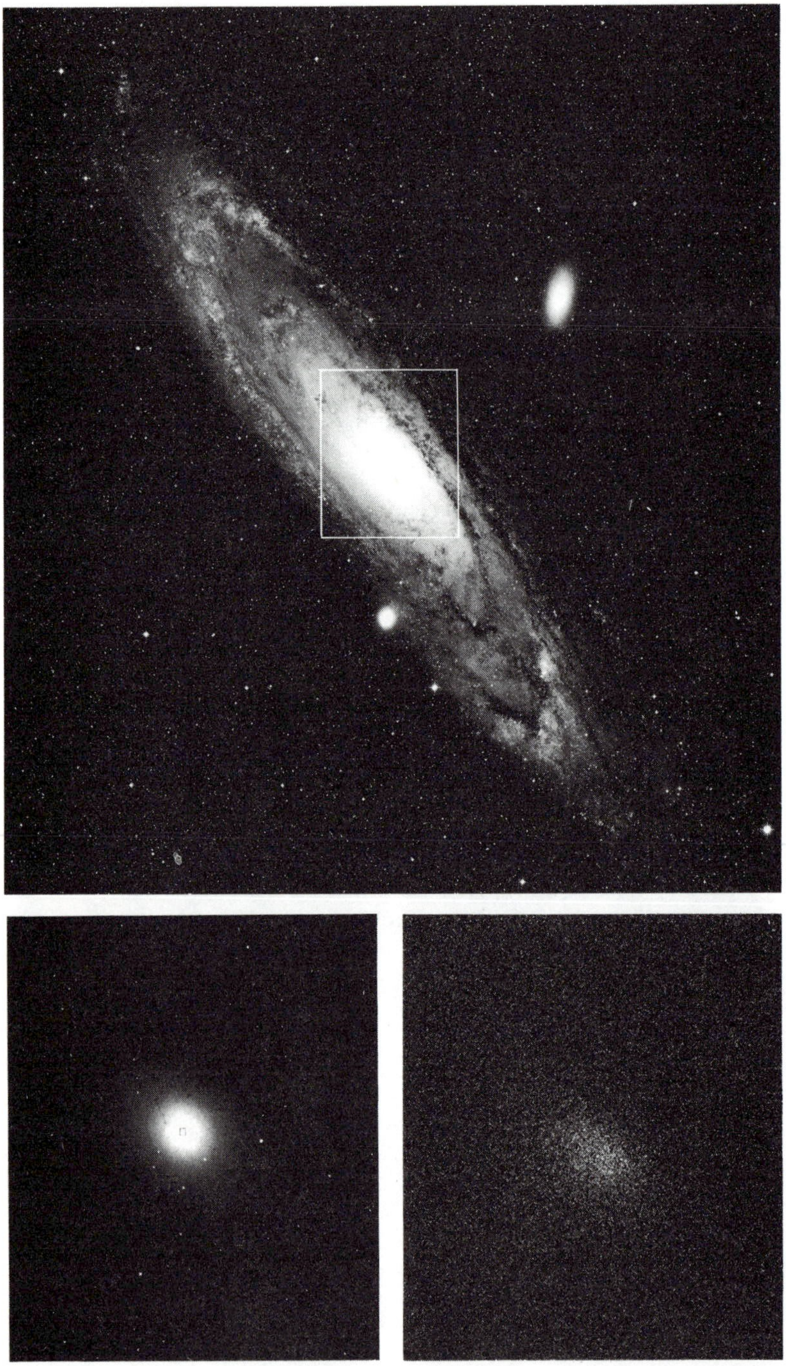

Figure 3.2 NUCLEUS OF THE GREAT NEBULA in Andromeda, or Messier 31 (M31), is 70 times farther away than the nucleus of our own galaxy. The central region of the top photo, which shows the entire galaxy, is heavily overexposed. The region within the rectangle is shown at bottom left. The photo at bottom right shows the central 20 parsecs of the nucleus, the area within the rectangle of the second picture. The white dots elsewhere are not stars but grains of photographic emulsion.

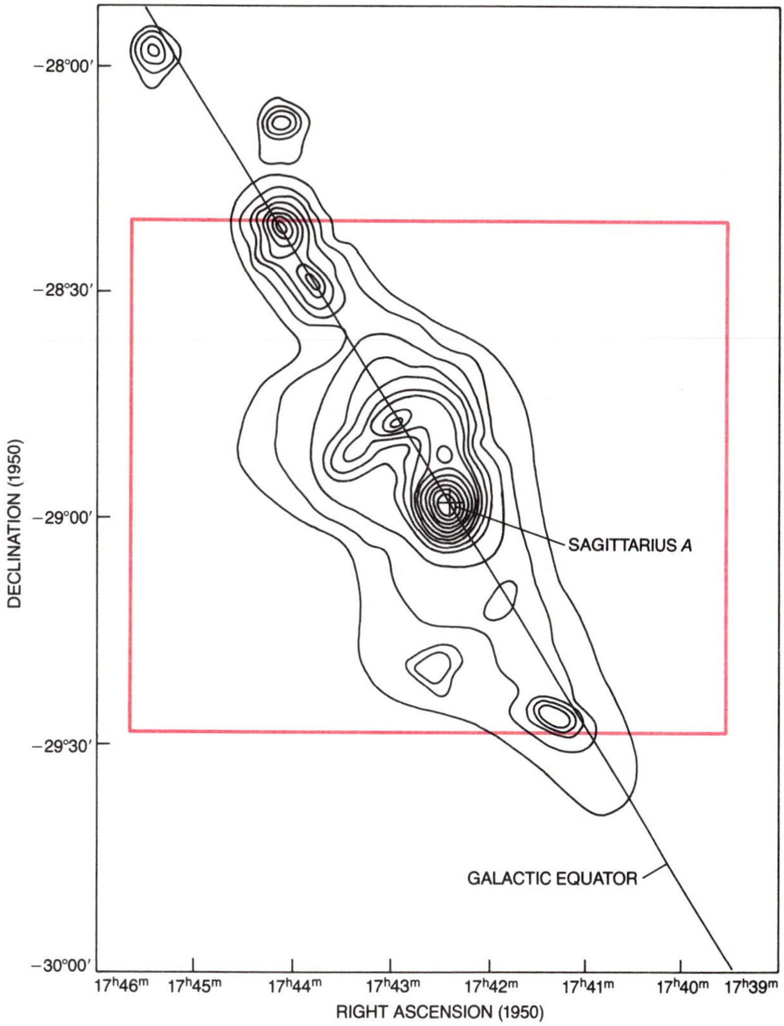

Figure 3.3 RADIO MAP OF THE GALACTIC CENTER was made at a wavelength of 3.75 centimeters by Dennis Downes and Arthur E. Maxwell of Harvard University and M. L. Meeks of the Massachusetts Institute of Technology with the 36.6-meter radio telescope at Westford, Mass. The principal source of the radio emission is ionized gas. The rectangle encloses the region represented in Figure 3.1. Ionized gas and stars are concentrated along galactic equator. They are brightest at the radio source Sagittarius *A*.

range of wavelengths when they are accelerated, for example during near-collisions with other charged particles (see Figure 3.4). In contrast, electrically neutral gas, such as the H I regions (which consist largely of hydrogen in atomic form) or molecular clouds (which consist largely of hydrogen in molecular form), emits little radio energy except at a few discrete wavelengths.

A different picture of the extended nuclear region emerges from observations made in the infrared at a wavelength of 2.2 microns, roughly four times longer than the visible wavelengths. Whereas the radio emission comes from hot interstellar gas, the 2.2-micron radiation is predominantly starlight. Most of the 2.2-micron radiation comes from red giants, which are large, luminous, highly evolved stars with surface temperatures as low as 2,000 degrees Kelvin, compared with 6,000 degrees K. for the sun. Although at most only a few percent of the stars in the galactic center are likely to be red giants, these objects should be fairly evenly distributed among the younger and brighter stars, and therefore in principle they should provide a good indication of the distribution of all stars in their neighborhood.

A color-coded picture of the extended central re-

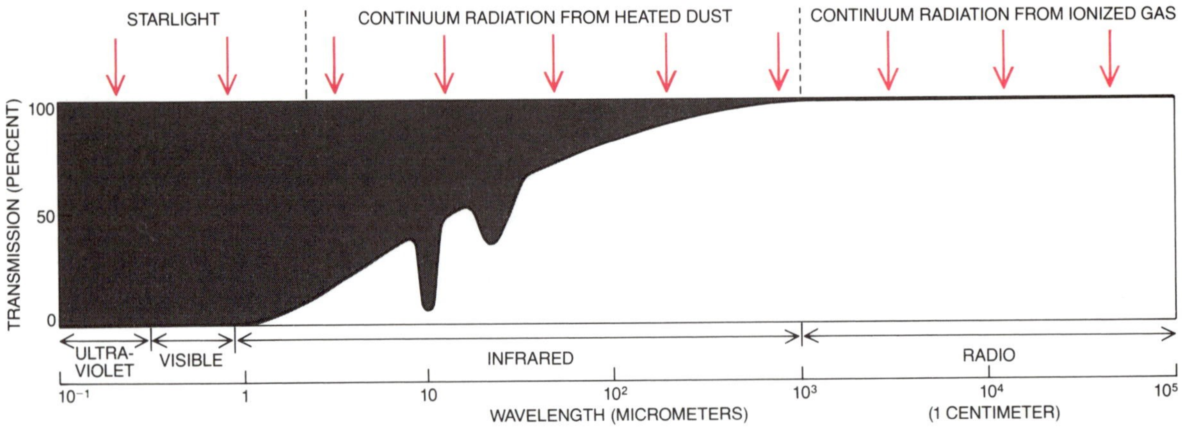

Figure 3.4 RADIATION SOURCES in the galactic nucleus include stars, heated dust and the clouds of ionized gas known as H II regions. The term H II describes hydrogen atoms that have been dissociated into protons and electrons by the ultraviolet radiation from nearby hot stars. The electrons and protons emit radiation over a broad band of wavelengths (the continuum radiation) as they are accelerated by electrical forces, primarily those they exert on one another. The curve shows how much continuum radiation survives passage through the interstellar dust to reach instruments on the earth. Before radio and infrared astronomy the galactic nucleus was inaccessible to observation.

gion of the galaxy at 2.2 microns appears in Figure 3.5. Some of the patchy appearance of the picture is accounted for by variations in the amount of interstellar dust between the solar system and different portions of the galactic center. The brightest point-like objects in the picture are probably red giants and are undoubtedly foreground objects. This inference is supported by their fairly uniform distribution across the picture.

Even with the foregoing exceptions it is clear that the 2.2-micron emission, like the radio emission, is concentrated along the galactic equator. The direction of maximum brightness at 2.2 microns coincides with a bright radio peak within Sagittarius *A* known as Sagittarius *A* West. Maps at longer infrared wavelengths, those emitted by the interstellar dust itself, also have their brightest peaks near Sagittarius *A* West. Hence the concentrations of stars, hot gas and cooler dust are all highest in the same direction. There can be little doubt that in this direction lies the actual center of the galaxy.

From the 2.2-micron observations, which were made by Eric E. Becklin and Gerry Neugebauer of the California Institute of Technology and the Hale Observatories, one can estimate the mass of stars at the center of the galaxy. Becklin and Neugebauer found that the strength and distribution of the 2.2-micron emission from the center of the Andromeda galaxy, designated M31, is comparable to the emission from the center of our galaxy. Since the nucleus of M31 can be clearly photographed at visible wavelengths, one can estimate with fair accuracy the number of stars that must be represented. By assuming that the ratio between infrared radiation and visible radiation is the same for the center of our galaxy as for M31, Robert H. Sanders and Thomas Lowinger of Columbia University and J. H. Oort of the University of Leiden calculated the distribution of stellar mass in the center of our galaxy. They estimate that a sphere with a diameter of one parsec centered on the bright central peak of the 2.2-micron emission contains a mass in stars equivalent to about two million suns.

The most densely packed globular cluster of stars outside the galactic nucleus has about 10,000 stars within its central parsec. Within one parsec of the sun there are no other stars. Assuming that each star in the central parsec is similar to our sun, the above estimate that the average distance between stars in the central parsec is about one light-week, or roughly 100 times the distance from the sun to the planet Pluto. If the earth were within the central parsec, the millions of nearby stars would bathe the planet with an amount of light equal to that of several hundred full moons. Whether life could

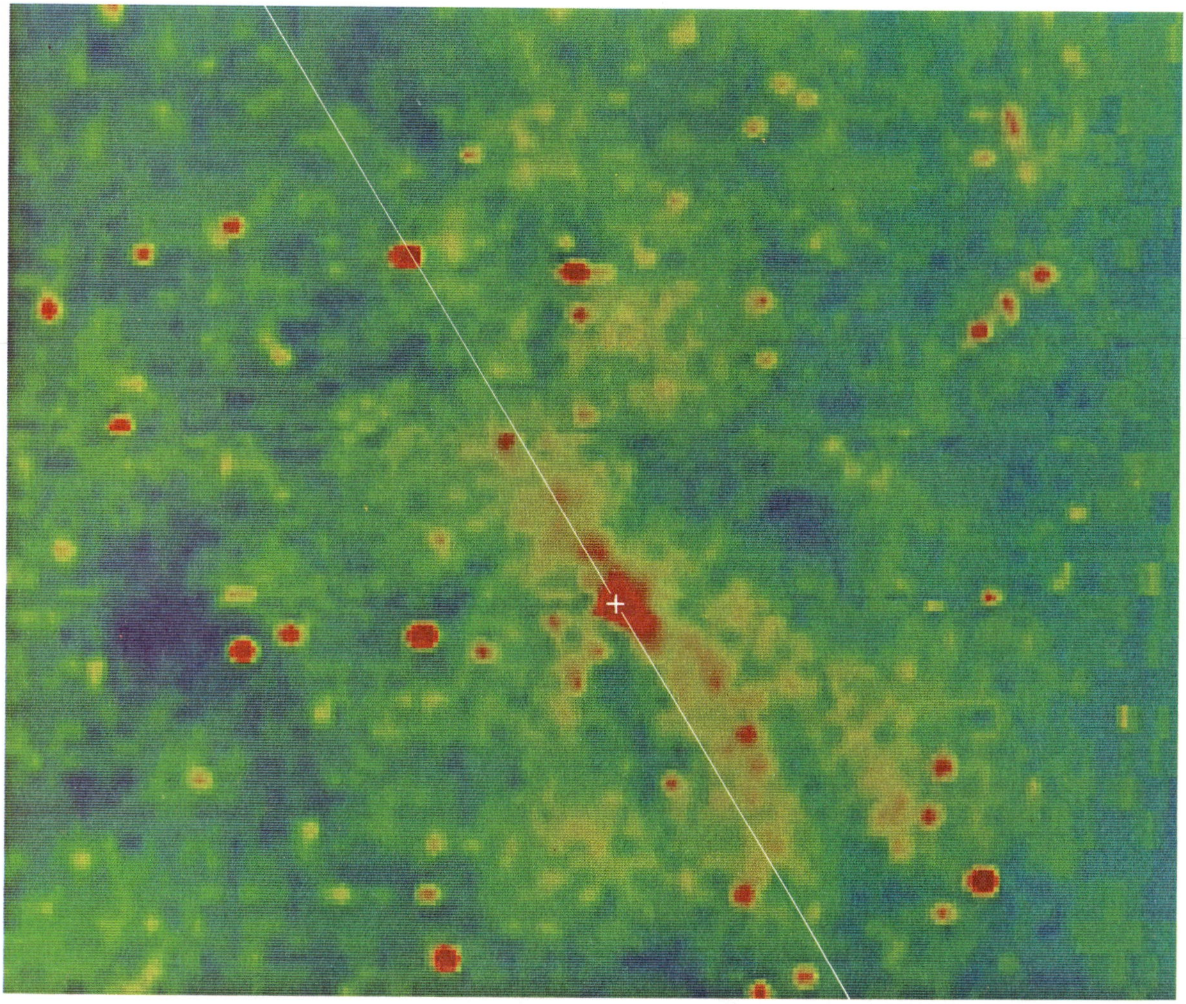

Figure 3.5 INFRARED VIEW OF THE GALACTIC CENTER was made at a wavelength of 2.2 microns by Eric E. Becklin and Gerry Neugebauer. Their observing instrument was the one-meter telescope at the Las Campanas Observatory in Chile. In this computer-generated color-coded picture the intensity of the radiation increases from blue to red. The cross marks the center of the galaxy. It can be seen that the infrared radiation tends to concentrate along the galactic equator, indicated by the thin white line. At 2.2 microns cool red stars are the principal source of radiation detected from the galactic center. Individual bright spots represent red-giant stars in the foreground. Patchiness is caused by interstellar dust.

exist within the radiation-filled central parsec is another question.

In observing a region as small and as distant as the central parsec of the galaxy it is desirable to work with instruments that have the highest possible angular resolution. The largest radio telescopes operating at short wavelengths are limited to a resolution of about one minute of arc, a resolution low enough to blur completely any spatial structure in the central parsec. By using two or more radio telescopes separated by a large distance and employing the technique of interferometry, however, the resolution can be increased substantially. Unfortunately the southerly location of the galactic center has made it almost inaccessible to interferometry with the large radio telescopes that until recently existed only at high northern latitudes. The situation will improve quite soon, however, with the completion next year of the Very Large Array radio facility near Socorro, N.M. This instrument will finally make it possible to explore the central parsec with an angular resolution comparable to that now attainable at infrared wavelengths.

The high resolution made possible at radio wavelengths by interferometry has already been applied in a limited way to the study of Sagittarius *A* West. In 1974 Bruce Balick, then of the Lick Observatory, and Robert L. Brown of the National Radio Astronomy Observatory in Green Bank, W.Va., discovered by radio interferometry that Sagittarius *A* West has near its center a radio-emitting object that is both exceedingly small and exceedingly bright. This compact radio source has since been shown to have a central core whose diameter is no larger than .001 arc-second, or barely 10 times the distance from the earth to the sun (assuming that the object is indeed at the center of the galaxy).

Although the compact radio object bears some resemblance to certain kinds of radio-emitting stars, it also meets some of the qualifications for the way matter should appear as it is drawn into a black hole. It is thought that a black hole may be created when a massive star that has exhausted its nuclear fuel can no longer generate the internal pressure needed to counterbalance the inward pull of gravity. Under such circumstances the interior of the star will collapse until it is so small and dense that known physical laws cannot describe it. The gravitational force near such a collapsed object is so great that there is a region surrounding it from which essentially nothing, not even light, can escape; the region is a black hole.

Material in the strong gravitational field just outside a black hole will be accelerated to high velocities and heated by collisions to very high temperatures. As a result a black hole should be surrounded by a small but intense disklike source of radio radiation, looking rather like the compact radio source. At the galactic center, where gas and dust are abundant, one would expect a black hole to swallow up matter at a comparatively high rate, growing rapidly in mass and in gravitational influence. The possibility that Sagittarius *A* West harbors a massive black hole at the center of the galaxy has undoubtedly entered the thoughts of many astronomers. I shall examine recent evidence for this possibility as we proceed.

The detailed examination of the central few parsecs of the galaxy has been much easier at infrared wavelengths than in the radio region. The study has been carried out at several infrared wavelengths over the past half-dozen years, principally by G. H. Rieke and Frank J. Low of the University of Arizona and by Becklin and Neugebauer. At each wavelength they have found a bright concentration of objects within the diameter of one parsec coincident with Sagittarius *A* West. At different wavelengths, however, the relative luminosity and distribution of the objects are strikingly different.

The observations at 2.2 microns and 10 microns perhaps best demonstrate the differences (see Figure 3.6). At each wavelength many distinct objects can be observed, but only six of them show up at both wavelengths. In other words, objects prominent at one wavelength are not prominent at the other. For example, the object known as IRS 1 (infrared source 1), which is the brightest object at 10 microns, is only moderately bright at 2.2 microns. IRS 7, by far the brightest object at 2.2 microns, is barely detectable at 10 microns. It would therefore seem that the objects being observed at different wavelengths belong to different classes.

What kinds of object are they? All are brighter in the infrared than the great majority of stars would appear to be at a distance of 10,000 parsecs. Since, as I have mentioned, red giants are particularly bright at 2.2 microns, it is natural to ask whether any of the bright 2.2-micron sources at the galactic center are red giants. There is a way to answer the question. In the atmospheres of the coolest stars molecules are able to form. One species that forms in abundance is carbon monoxide, which absorbs radiation beginning at 2.3 microns and extending to longer wavelengths. The absorption occurs most strongly in luminous red giants. In 1975 observations made by groups at the University of Arizona and at Cal Tech revealed absorption dips at 2.3 microns in the spectrum of three of the galactic-center sources (IRS 7, IRS 11 and IRS 12), showing that each source consists of at least one luminous red giant. On the other hand, one of the sources that is bright at 2.2 microns, IRS 16, shows little or no carbon monoxide absorption (see Figure 3.7). The identification of IRS 16 is still open to question. Its position seems to coincide with the compact radio source in Sagittarius *A* West, so that it is of particular interest.

A second object that does not show carbon mon-

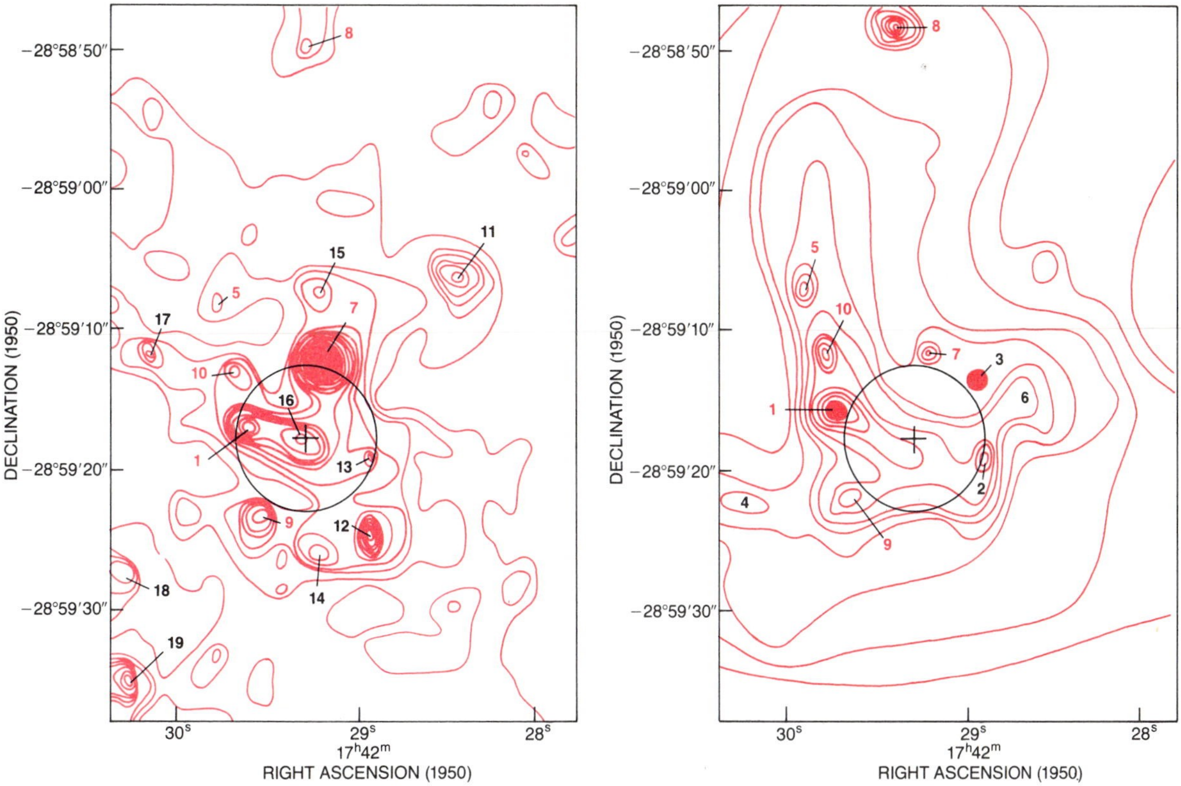

Figure 3.6 INFRARED MAPS of the galactic center plotted at 2.2 microns (*left*) and 10.6 microns (*right*). The circle on each map represents what is thought to be the central parsec of the galaxy. IRS 7, IRS 11 and IRS 12 are red giants. Except for IRS 3 and IRS 7 most of the sources on the 10.6-micron map are thought to be warm dust clouds associated with H II regions. The cross on both maps coincides with the bright radio-emitting object within Sagittarius *A* West. The 2.2-micron map was made by Becklin and Neugebauer with the 200-inch telescope. The 10.6-micron map is a composite of several maps obtained with various telescopes by G. H. Rieke of the University of Arizona and C. M. Telesco and D. A. Harper of the University of Chicago.

oxide absorption, IRS 1, belongs to the class of objects that are much brighter at 10 microns than at 2.2. Very little of the 10-micron emission from objects of this class can be direct stellar radiation because even the coolest stars are much fainter at 10 microns than at 2.2. Infrared astronomers have now found many sources scattered around the galaxy that are unusually bright at 10 microns. In almost every case the source of the radiation has been identified as warm dust. The dust particles absorb visible and ultraviolet radiation from nearby stars or other luminous objects and reradiate the absorbed energy at infrared wavelengths. Something similar probably happens in the galactic nucleus, and one can assume that the peaks in the map at 10 microns represent dust clouds in or near the central parsec.

The interstellar gas in our galaxy is about 100 times more abundant by mass than the interstellar dust, and its presence is always implied by the presence of dust. It therefore seems reasonable that the radio-emitting gas of Sagittarius *A* West is gas associated with dust clouds and that it is heated by the same source or sources warming them. If this is the case, one would expect high-resolution radio maps of Sagittarius *A* West to reveal structure similar to that seen in the maps at 10 microns. There were indications in 1974 from radio-interferometry studies that Sagittarius *A* West is indeed a collection of

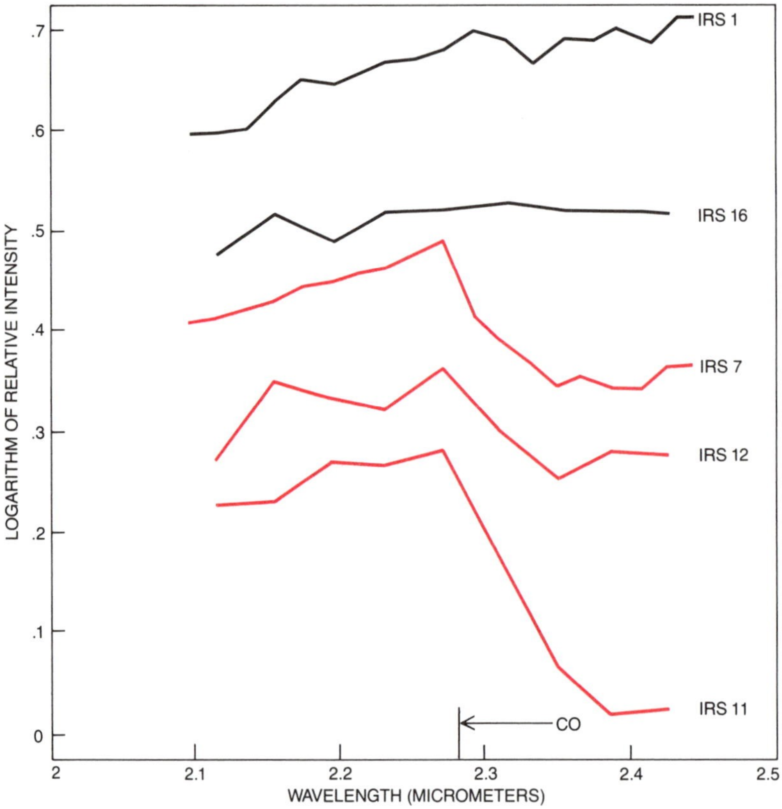

Figure 3.7 INFRARED SPECTRA OF FIVE SOURCES close to the galactic center show that three, IRS 7, IRS 11 and IRS 12, carry the "signature" of red giants: a drop in intensity at wavelengths longer than 2.3 microns. Such stars have an atmosphere rich in carbon monoxide, which strongly absorbs radiation at 2.3 microns and beyond. IRS 1, an H II region, and IRS 16, in Sagittarius *A* West, do not exhibit such absorption. The spectra were obtained with the 200-inch telescope by Becklin, Neugebauer, Steven V. W. Beckwith, Keith Matthews and C. G. Wynn-Williams.

more compact clouds of ionized gas. At the time, however, it was difficult to determine whether or not any of the emission peaks in the infrared and radio maps were spatially associated.

The story up to this point approximately covers the state of knowledge about the galactic nucleus in 1975. It was known that within the central parsec there was a massive concentration of stars whose visible light is hidden by dust. Within the central few parsecs there had been found infrared-emitting sources of several types. Of the sources that were bright at 2.2 microns some had been shown to be cool and luminous red giants; most of the other sources might also harbor cool stars but they had not yet been identified. The majority of the objects that were bright at 10 microns resembled compact H II regions, but although it was known that ionized gas was present at the galactic center, it was not yet known whether the gas was physically associated with any of these infrared sources. An intriguing source of radio emission, which might be a supermassive object at the very center of the galaxy, had been found to coincide with a peculiar bright source that emits strongly at 2.2 microns. Finally, nothing whatever was known about the motions of all these objects; the material in the central parsec might be collapsing, expanding, rotating or moving randomly.

The stars, gas and dust at the galactic center emit

radiation whose variation in intensity with respect to wavelength is fairly smooth; the radiation is therefore called continuum radiation. Another potentially rich source of information is line radiation: the radiation emitted and absorbed at specific and characteristic frequencies by atoms and molecules. The spectrum of an astronomical object usually consists of both a smooth continuum and lines. Each line is associated with a specific energy level of an atom or a molecule and has an intensity, a central wavelength and a width.

The intensity of the line is a measure of the temperature and the abundance of the emitting substance. If the source is moving toward the observer, the line is shifted to a shorter wavelength; if the source is receding, the line is shifted to a longer wavelength. Such Doppler shifts inform the astronomer about radial velocity: the velocity of a source along the line of sight. Velocities toward the observer ("blue shifts") are termed negative; those away from the observer ("red shifts") are termed positive. When the source atoms or molecules are moving at varied speeds, the shifts vary. The line is "smeared out" and is said to have a width. Therefore the width of a line can reveal the range of radial velocities in the source.

In 1973 David Aitken, Barbara Jones and James Penman of University College London succeeded in detecting for the first time an infrared emission line originating from the galactic center. The line's wavelength near 12.8 microns identified its source as being singly ionized neon, or Ne II: neon missing one of its 10 electrons. The same neon line had earlier been found to be one of the brightest infrared emission lines in several ionized regions elsewhere in the galaxy. These regions also reveal themselves by continuum radiation in the radio region of the spectrum.

Using the known intensity of radio radiation from Sagittarius *A* West and assuming that the abundance of neon in the ionized gas of Sagittarius *A* West is normal, Aitken, Jones and Penman calculated the expected intensity of the neon line. The observed intensity of the line was slightly greater than their predicted intensity, which implied that the gas at the galactic center is richer in neon than gas observed elsewhere. The prediction was close enough, however, to imply that the same ionized gas produces both the infrared neon line and the radio continuum emission. Hence the London work showed that the 12.8-micron line of Ne II, representing only a tiny fraction (approximately .01 percent) of the ionized gas, could serve as a tracer in studying the motion and distribution of the ionized gas of Sagittarius *A* West, which was putatively at the very center of the galaxy.

The London group worked with an infrared instrument that admitted a circular beam whose diameter corresponded roughly to one parsec at the galactic center. The resolution of their spectrometer was too low to determine the width of the neon line or whether it exhibited a Doppler shift. In the department of physics at the University of California at Berkeley, where I was then a graduate student, a succession of students working with Charles H. Townes had been developing infrared spectrometers capable of high spectral resolution. In 1975 Eric Wollman, John Lacy and I, together with Townes and David Rank of the University of California, Santa Cruz, measured the shape of the Ne II line with the 1.5-meter telescope of the Cerro Tololo Inter-American Observatory in Chile (where Sagittarius passes directly overhead). What we found made it clear that Sagittarius *A* West is not a typical galactic H II region. The neon line had a width of several hundred kilometers per second, whereas for an H II region velocities in the range of 30 kilometers per second are normal (see Figure 3.8).

At about the same time radio observations of emission lines generated by the recombination of hydrogen nuclei and electrons made by Thomas Pauls, Peter G. Mezger and Edward Churchwell of the Max Planck Institute for Radio Astronomy in Bonn also showed that the ionized gas in Sagittarius *A* West was in rapid motion. Whether the motion was organized or turbulent and how the neon and other ionized gas was distributed within the central parsec had not yet been determined.

The answers to these and other questions called for observations of the 12.8-micron line on a finer spatial scale. At the Lick Observatory in 1975 and 1976 we observed several positions within the central parsec and found that both the line strength and the radial velocity varied with position. In 1977 Lacy completed a new and more sensitive spectrometer that made it possible to observe the Ne II in even greater detail. In 1977 and 1978 Lacy, Townes, Fred Baas (visiting from Leiden) and I, working with this instrument at the new 2.5-meter Du Pont telescope at the Las Campanas Observatory in Chile, got complete coverage of the Ne II line over the

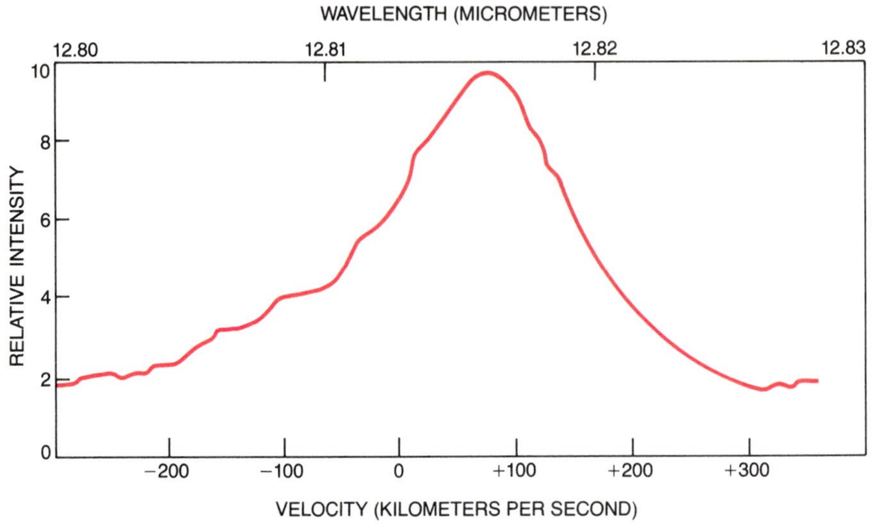

Figure 3.8 EMISSION OF SINGLY IONIZED NEON, Ne II, present in gas clouds in the central parsec of the galaxy. Examination of the Ne II emission line near 12.8 microns reveals that the line has been substantially broadened by motion of the gas clouds in which the neon is a minor constituent. Motion toward the observer (designated as negative) shifts the line to a higher frequency, or shorter wavelength, whereas motion away from the observer (designated as positive) has the opposite effect. The intensity of continuum radiation near 12.8 microns from the warm dust in the nucleus is given by the level of the radiation on each side of the broadened neon line.

central parsec and beyond it. Working with aperture diameters corresponding to as little as .2 parsec at the galactic center, we obtained spectra on a rectangular grid of positions and made contour maps of the line intensity at several different velocities.

Our data show that there is great diversity in both the intensity of the neon line and the velocity of the neon observed in the direction of the central parsec. In a few directions the intensity of the line exhibits peaks at more than one velocity (see Figure 3.9). Sagittarius *A* West clearly cannot be considered a uniform region of ionized gas but rather must be composed of a number of independent smaller sources moving with respect to one another. As I have mentioned, much of the infrared continuum radiation at 2.2 and 10 microns also comes from groups of small sources. It was important to determine whether the distribution of ionized neon was related to either the 2.2- or the 10-micron maps.

In order to investigate this question we noted the positions of maximum intensity of the Ne II line at specific velocities. When we plotted the positions of the velocity components on the infrared-continuum maps, we found that the maximum intensities coincided with many of the sources that were bright at 10 microns. For example, IRS 1 is at the same position as a peak in the neon line at zero velocity. Near IRS 2 there is little zero-velocity ionized neon, but there is a sharp peak in the ionized neon approaching the observer at about 280 kilometers per second, a velocity not seen elsewhere in the central parsec. At least five other infrared sources, IRS 4, IRS 5, IRS 6, IRS 9 and IRS 10, coincide with localized velocity components of the Ne II line (see Figure 3.10).

The spatial coincidences of hot ionized gas and warm dust are convincing evidence that most of the bright sources on the 10-micron continuum map are compact H II regions with Sagittarius *A* West. From the intensity of the neon line we can estimate that the ionized material in each of these regions amounts roughly to one solar mass. The mass of the associated dust, which is responsible for the 10-micron continuum emission, is much less than that. The diameter of a compact H II region is typically a quarter of a parsec. Since there are about two million stars in the central parsec, it is likely that each compact H II region harbors as many as 100,000 stars.

Only two of the sources on the 10-micron map are definitely not H II regions. One of them, IRS 7, is

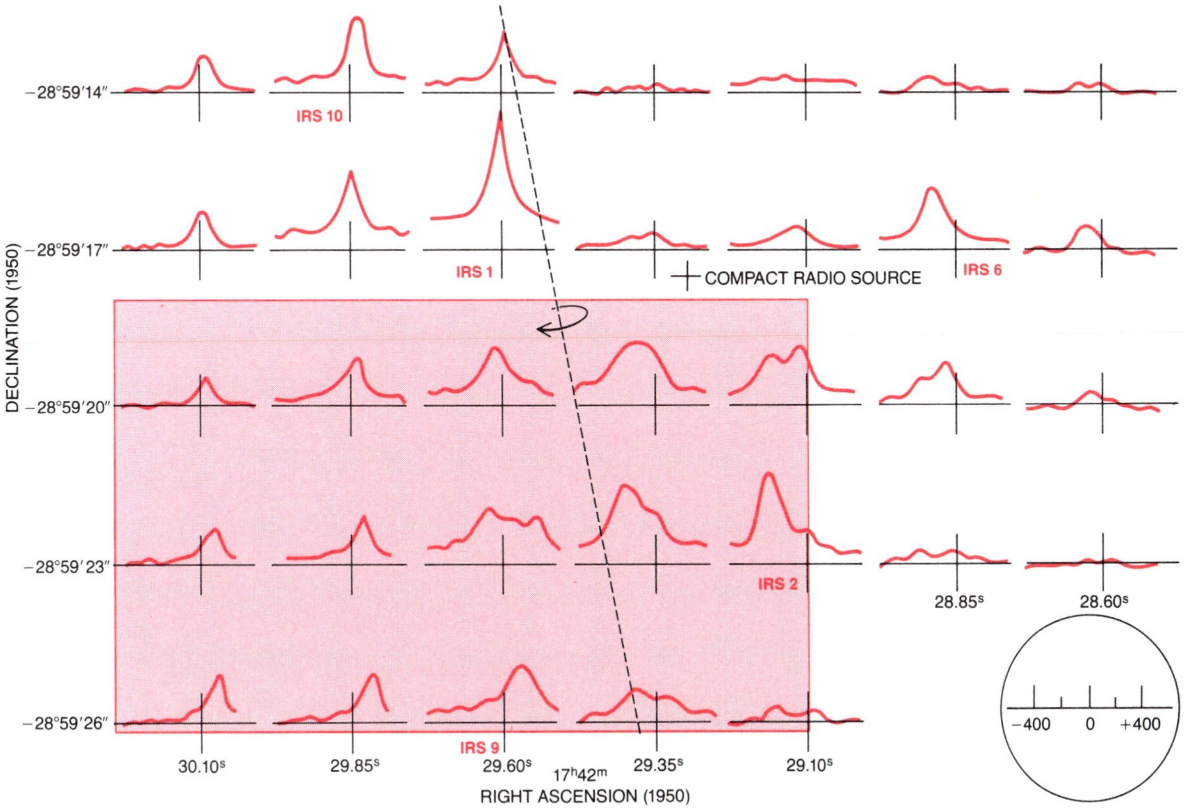

Figure 3.9 DETAILED STUDY OF THE NEON LINE. The grid of spectra is arranged according to position in the sky and includes slightly more than the central parsec. A circle (*lower right*) gives the beam size of the spectrometer; a scale gives the velocity of gas motion in kilometers per second. Several of the localized velocity components of the neon line coincide with sources that are bright at 10 microns (IRS 1, IRS 2, IRS 6, IRS 9 and IRS 10), indicating that the ionized gas is physically associated with warm dust clouds. The sloping broken line separates regions of ionized neon that are predominantly approaching (*right*) and receding (*left*). Such a well-defined separation suggests that material in the central parsec is rotating in the sense that is given by the arrow. The region within the rectangle reappears in Figure 3.10.

already known to be a red giant. The other, IRS 3, is more mysterious. IRS 3 cannot be a red giant because it is not bright at the shorter infrared wavelengths. Nor can it be an H II region because there is a distinct drop in the intensity of the Ne II line in the vicinity of the source. This suggests that IRS 3 is a cool and opaque cloud of dust and gas, either recently ejected by a star or collapsing gravitationally to become a star, in front of the central parsec.

Each of the compact H II regions in the central parsec is moving at a velocity whose radial component is given by the Doppler shift of its associated neon line. Most of the H II regions are well separated in space or in velocity and therefore must be independent objects. Nevertheless, study of all the neon spectra indicates that there is a collective motion of the central parsec. This is seen most easily by noting that a line running roughly north-south through the middle of the region separates the ionized neon that is predominantly approaching from the ionized neon that is predominantly receding. The simplest interpretation of the division of velocities is rotation; the dividing line would then correspond to the axis of rotation.

Although evidence for rotation has been found so far only in the ionized gas, it seems reasonable that the rest of the material in the central parsec (includ-

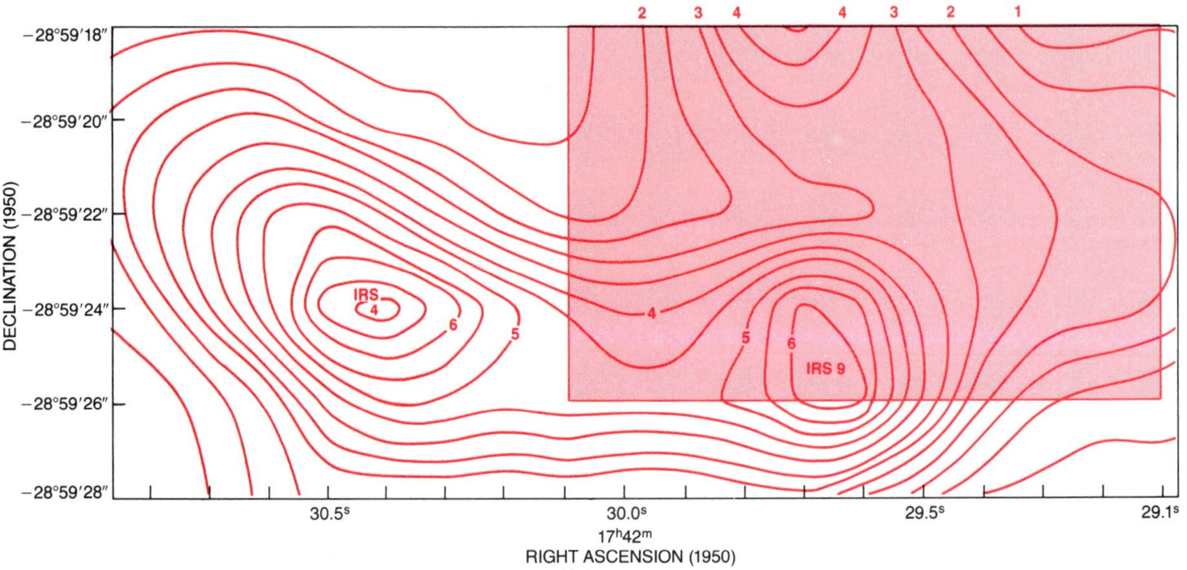

Figure 3.10 INTENSITY OF THE NEON LINE in the southeastern part of the central parsec demonstrates the association of the ionized neon with two specific sources, IRS 4 and IRS 9, whose continuum radiation originates in warm dust clouds. The contour lines show the intensity of neon with velocities of about +125 kilometers per second. The peaks coincide with the two infrared sources. Spectra made at the positions of the sources show that IRS 4 is actually receding at 110 kilometers per second and IRS 9 at 140 kilometers per second. Observations for this figure and Figure 3.9 were made by the author and colleagues with 2.5-meter telescope at Las Campanas Observatory.

ing the stars) is moving in much the same way as the ionized gas. On the basis of the Doppler shifts of the neon line observed a few tenths of a parsec from the apparent axis one can estimate that the time required for gas, dust and stars inside the central parsec to make one trip around the galactic center is roughly 10,000 years. (The sun takes about 200 million years.) The orientation of the axis is peculiar: it is almost perpendicular to the axis of rotation of the main galactic disk. This suggests that the nuclear region is a separate structure, distinct from its surroundings. It is not yet known how far out from the center the peculiar axis of rotation extends.

The motions of the ionized gas in the central parsec, whether or not they are part of a collective rotation, are influenced by the gravitational force of the mass within the same volume and can be utilized to calculate that mass. Since the rest of the galaxy is distributed more or less symmetrically around the center, with most of its mass a great distance from the center, it exerts almost no net gravitational force on the central parsec. The total mass of the central parsec, as estimated from the neon-line velocities, is between five and eight million times the mass of the sun. Of that total it appears from study of the 2.2-micron observations that only two million solar masses are in the form of stars of all types. The total mass of dust and gas within the region is in comparison negligible: probably no more than about 10 solar masses.

It is therefore fascinating to speculate whether the difference of from three million to six million solar masses between the apparent stellar mass and the total mass is significant. Does it represent missing mass still to be accounted for? Or are the estimates inaccurate? One observation may be pertinent. Measurements of the infrared neon line and of the radio lines generated by the recombination of hydrogen nuclei and electrons show that within the central few parsecs the gas velocities tend to increase toward the center. It has been pointed out by Luis Rodriguez and Eric J. Chaisson of the Center for Astrophysics of the Harvard College Observatory and the Smithsonian Astrophysical Observatory, and by others as well, that such an increase could not be due to the gravitational force of the central stellar cluster unless the cluster had an unusually dense core or possibly included a compact, massive and presumably nonstellar central object.

In considering theoretical models to account for the observations of the central parsec let us assume (at least initially) that the nucleus is not being observed at a particular epoch. The observations have shown that the central parsec contains an extremely dense cluster of stars, together with a sizable amount of dust and ionized gas, which are distributed nonuniformly and are in rapid motion. A satisfactory model should be able to explain not only the presence of such features but also their continued presence.

Because of the gravitational attraction toward the center of the galaxy it is not surprising to find the center populated by a dense cluster of stars. Occasional newcomers to the cluster are probably ordinary stars that had previously circled the galactic nucleus at some greater distance and that have lost angular momentum, perhaps as a result of near-collisions with other stars. Conversely, through the same process stars within the nucleus may occasionally acquire enough velocity to escape from it. Hence the stellar population of the central parsec probably remains close to a steady state or at most changes very slowly.

In contrast, dense concentrations of dust and ionized gas are normally shortlived. In the outer parts of the galaxy there are two classes of such objects: the H II regions found in the vicinity of newly formed hot, luminous stars, and the planetary nebulas, which are also H II regions but whose ionization is due to certain old stars that are usually hotter but less luminous. In fact, ultraviolet radiation from roughly 100 stars of the type that ionize planetary nebulas would be needed to account for the amount of ionized gas observed in the central parsec, but as we have seen only about 10 ionized clouds are observed there. An identification of the clouds as planetary nebulas can be questioned for this reason. On the other hand, only about 10 highly luminous, newly formed stars of the type that ionize normal H II regions would be needed to do the job. Other typical sources of dust and gas, such as the "wind" of particles emitted by stars or the infall of matter from surrounding regions, would not give rise to the highly nonuniform distribution of interstellar material inferred from the infrared-continuum and neon-line observations.

In a region as densely populated with stars as the central parsec is, however, an additional mechanism for producing dense clouds must be considered: stellar collisions. A disruptive collision, one in which two stars actually make contact, will tear material out of one or both of them, creating a turbulent and initially dense cloud that disperses into the more rarefied surroundings. Calculations by Townes, Lacy and David Hollenbach indicate that within the central parsec collisions in which at least one participant is a red giant are an important source of interstellar material. The clouds that would be created by such collisions have sizes, velocities, internal motions and lifetimes that are consistent with the properties of the clouds actually observed. Moreover, the estimated collision rate may be adequate to yield clouds in the appropriate number.

The next question is: What is the source of the ionizing radiation needed to convert the clouds in the central parsec into H II regions? I have mentioned hot, luminous stars and planetary nebulas as possible ionizing agents. Hot, luminous stars remain that way for only a few million years, and so they cannot have traveled far from where they are now observed, which is in the galactic nucleus itself. Star formation, however, is thought to be a delicately balanced process involving the contraction of large clouds of gas and dust over a period of roughly a million years. In the turbulent environment of the galactic nucleus, where orbital periods may be as short as 10,000 years, it may not be possible for large volumes of gas and dust to contract into stars before they are thoroughly disrupted.

The most important constraint on the type of ionization source in the central parsec appears to be the uniformly low level of ionization in the region. Typical H II regions elsewhere in the galaxy usually contain, in addition to singly ionized neon, atomic species that are multiply ionized, that is, lacking two or more electrons. In order to create the more highly ionized states very energetic ultraviolet photons from extremely hot stars are needed.

In the course of our observations of Ne II we also searched intensively within the central parsec for spectral lines of doubly ionized argon (Ar III) and triply ionized sulfur (S IV) at their known infrared wavelengths. These lines are often quite strong in H II regions outside the galactic nucleus. The results at various positions in the central parsec were negative, with the exception of a very weak Ar III line detected in the direction of IRS 1. Recently a strong infrared line of Ar II (singly ionized argon) from the central parsec was detected by Steven P. Willner, Ray Russell and Richard Pueter of the University of California, San Diego, Thomas Soifer of Cal Tech and Paul Harvey of the University of Arizona. Their result implies a moderate overabundance of argon in the central parsec and, when it is combined with our result, makes it clear that in the central parsec the ionization state is unusually low.

The central stars of planetary nebulas, which are usually very hot, would seem to be ruled out as the source of ionizing radiation because most of them give rise to high ionization states. The situation is also peculiar, however, if several hot, luminous, recently formed stars are the principal source. Unless the most massive and hottest stars are somehow inhibited from forming at the galactic center, or unless those that formed most recently have cooled significantly, one would expect the ionizing stars to exhibit a wide range of temperatures. Such a temperature range would lead to variations in the ionization state of the gas throughout the central parsec. The absence of Ar III and S IV throughout the entire central parsec seems to imply either that all the ionizing stars are nearly identical or that there is one extremely luminous but relatively low-temperature ionizing source in the central parsec. An object of the latter type could not be an ordinary star.

Such considerations lead to a rather exotic model of the central parsec: a model dominated by a massive black hole at the center (see Figure 3.11). Material being swallowed up by a black hole is thought to be first stored outside it in a rotating disk, which under the right conditions is expected to radiate strongly in the ultraviolet. The uniformity of the ionization state of the central parsec might be ex-

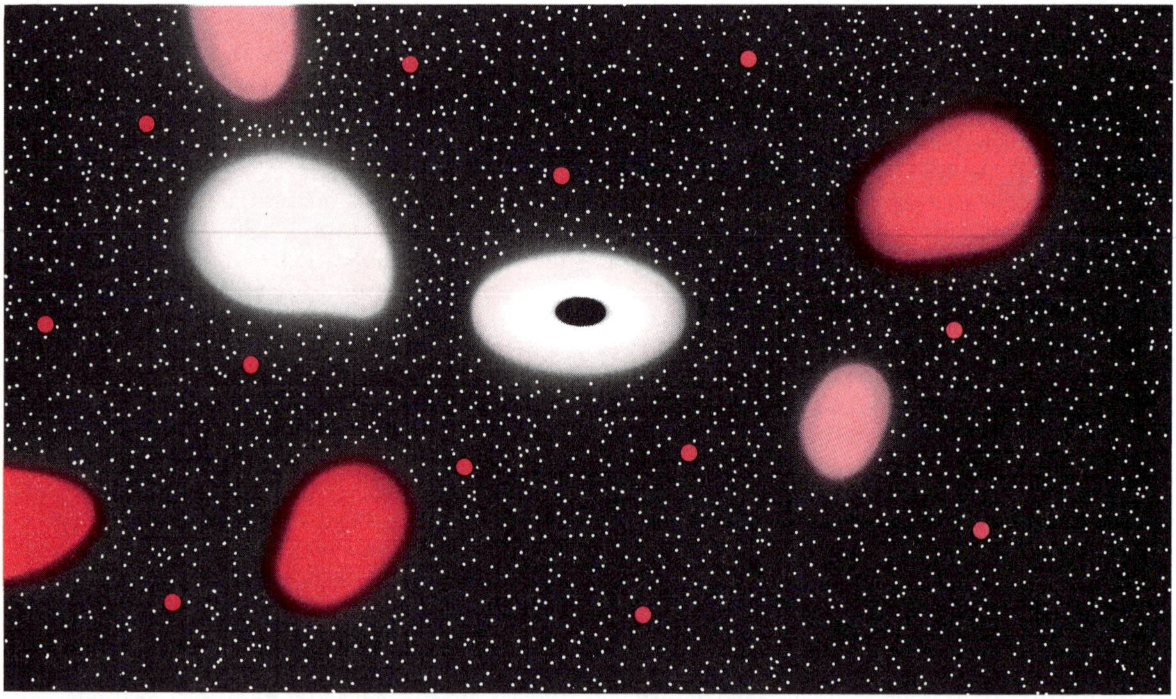

Figure 3.11 SCHEMATIC VIEW OF GALACTIC CORE postulates the existence of a massive black hole (*shown much larger than scale*) at the very center. The luminous color patches represent clouds of interstellar gas and dust, some of which form a hot, rotating accretion disk before it is swallowed and disappears. Intense ultraviolet radiation from the accretion disk ionizes the gas in the surrounding clouds, which then emit the infrared lines and the radio continuum radiation detected from the direction of the central parsec. Dust in the clouds is also heated by radiation from the accretion disk and nearby stars and emits infrared continuum radiation. Much of the short-wavelength infrared radiation is emitted by a few red giants (*colored dots*). The white dots are normal stars.

plained by the presence of such an object. A natural candidate for the black hole is the compact radio object, with a diameter of less than .001 arc-second, in Sagittarius *A* West.

The expected properties of massive black holes have been calculated by many theoretical astrophysicists. The spectrum and the intensity of the ionizing radiation from the disk depend only on the mass and the accretion rate of the black hole. Townes, Lacy and Hollenbach have shown that quantitative agreement with the observed luminosity and ionization state of the H II region in the central parsec of the galaxy would be provided by a black hole that had a mass of a few million suns and accreted matter at the moderate rate of about 10^{-5} solar mass per year. The necessary mass of accreting material could be supplied many times over by infall, stellar winds, planetary nebulas and the debris of stellar collisions within the central parsec. Even allowing for some of the material's being blown away by the intense radiation pressure from the accretion disk, from perhaps two million stars and from other luminous objects in the central parsec, there would seem to be ample material to feed the black hole at the required rate.

It is perhaps significant that the intensity of infrared radiation emitted by the peculiar object IRS 16, in which the compact radio source appears to be located, is roughly the same as the intensity that would be emitted in the vicinity of a black hole accreting matter at the above rate. Finally, the existence of a black hole with a mass of a few million suns is consistent with the difference between the estimated total mass in the central parsec and the mass of the stars in the same volume. A black hole of that mass fits easily into a volume the size of the compact radio source.

The evidence for the presence of a massive black hole appears reasonable, but it is indirect. More direct tests might come from future radio and infrared observations of the small region around the compact radio sources. For example, the intense gravitational field in the vicinity of such a massive black hole should give spectral lines originating in the source a Doppler width of 1,000 kilometers per second or more. When our infrared spectrometer with a beam diameter of .2 parsec was centered on the compact radio source, a possible location of the black hole, the neon line did not appear to be that wide. A weak and broad component that could not have been detected by our instrument may nonetheless have been present. Significant changes in the accretion rate of the black hole, which may occur on a time scale of only a few years, should eventually give rise to observable changes in the ionization state of the central parsec. Hence continued monitoring of all the infrared emission lines is important.

One might well ask again: Are we really observing the central parsec of the galaxy? In my view the answer is a strong yes. First of all, we are certainly pointing our telescopes in the direction of the central parsec; that is indicated by all the infrared and radio observations, perhaps most clearly by the 2.2-micron observations of the stellar distribution. Second, the rich array of infrared sources observed in this direction has few if any counterparts elsewhere in the galaxy. The high positive and negative velocities obtained from the neon lines are unique, and the enormous mass densities implied by the 2.2-micron and neon-line measurements have not been found elsewhere.

It is possible that the central parsec is the source of phenomena that are observed far beyond the galactic nucleus. Sudden large increases in the accretion rate of the proposed black hole, supernovas or bursts of star formation near the galactic center may be the cause of the outward-moving molecular clouds that have been observed by radio astronomers. Such outbursts would eject material from the galactic nucleus in the form of high-velocity clouds. The material eventually would collide with and accelerate the observed molecular clouds, which in turn could be the debris of earlier explosions.

Although the nucleus of our own galaxy is a powerful energy source, it is quite weak compared with many other galactic nuclei. Perhaps that is one reason we are here and able to study the galactic nucleus. At present it may be going through a quiet phase between more explosive epochs. The continued study of the galactic nucleus will lead to a better understanding of it and perhaps also of physical events in the nuclei of more active galaxies.

Globular Clusters

4

They are dense crowds of ancient stars bound together by their own gravitation. For decades the study of clusters has yielded insights into the evolution of stars, of galaxies and of the universe as a whole.

• • •

Ivan R. King
June, 1985

To a person looking through a large telescope a globular cluster is one of the most beautiful objects in the sky. Stars fill the field of view by the thousands; as many as a million, most of them too faint to be visible, may be packed into a spherical space whose diameter is typically less than 150 light-years. For decades astronomers have pondered how such crowds of stars may have been formed and how the interaction of their gravitational fields holds them together in a stable cluster.

Throughout the 20th century, moreover, the study of globular clusters has led to fundamental advances in many branches of astronomy. In part because the clusters are so luminous their spatial distribution has helped investigators to stake out the frontiers of the Milky Way and of other galaxies; it has even been suggested that the clouds of gas that engendered globular clusters were the building blocks from which galaxies were made. Furthermore, all stars in a given cluster can be assumed to be the same age, and so the types of stars in clusters offer general insight into how stars evolve and why they differ in color and brightness. Finally, globular clusters bear on the evolution of the universe itself. They are the oldest objects known, dating perhaps from just after the big bang; as a result their ages impose severe observational constraints on cosmological models, and their chemical composition is evidence of the composition of galaxies at the earliest stage of development.

At a time when each of these subjects—galactic structure, stellar evolution and classification, and cosmology—has become the concern of a quasi-independent discipline, the study of globular clusters still conveys, in addition to its intrinsic interest, a sense of the underlying unity of astronomy. Indeed, if astronomers could answer all the questions about globular clusters that continue to perplex them, they would know a great deal more than they do now about the nature of the universe.

GALACTIC STRUCTURE

At the turn of the century our stellar system was thought to consist of a disk only a couple of thousand parsecs in diameter centered on the sun. (One parsec is 3.26 light-years.) This heliocentric conception came into serious question in 1918, when Harlow Shapley used the telescopes of the Mount Wilson Observatory to measure the distance to several

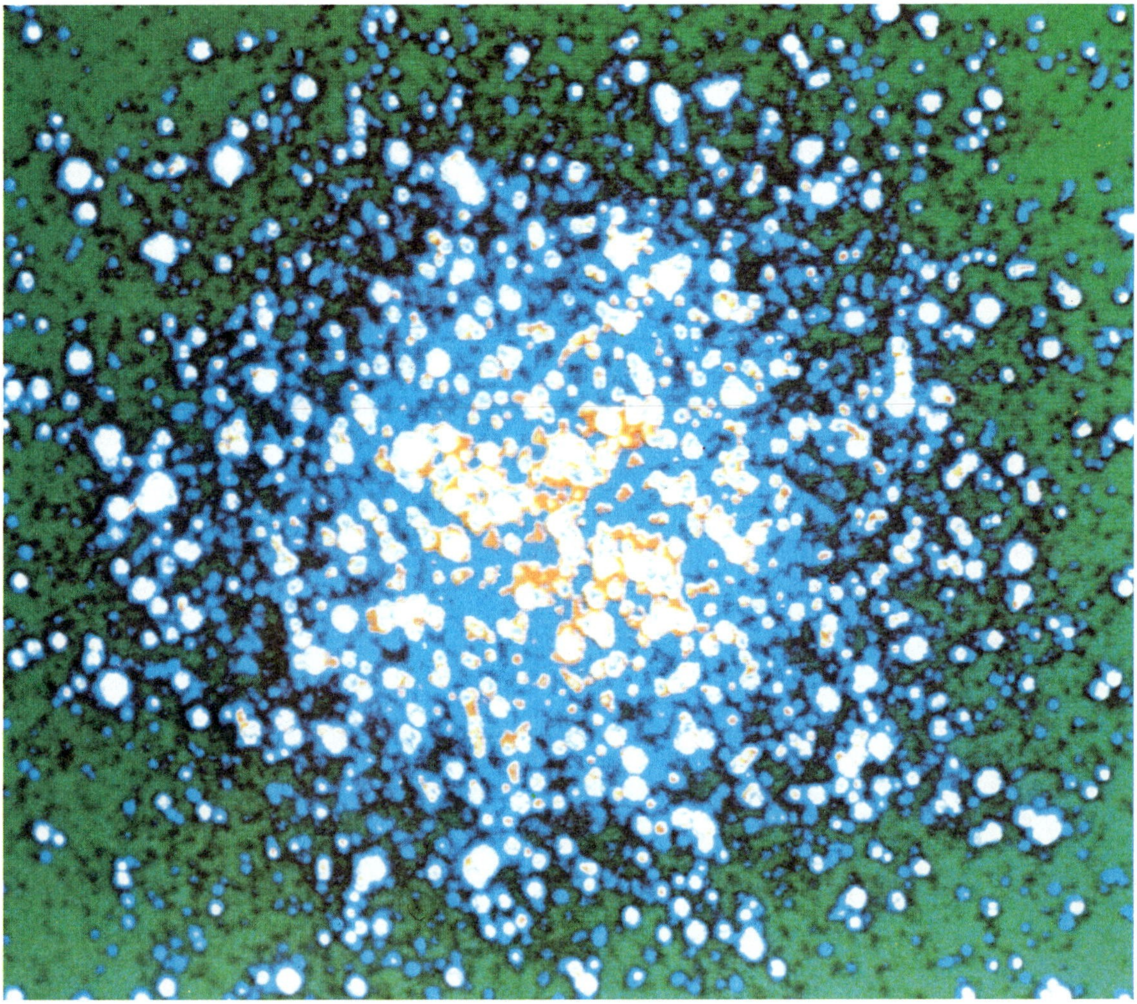

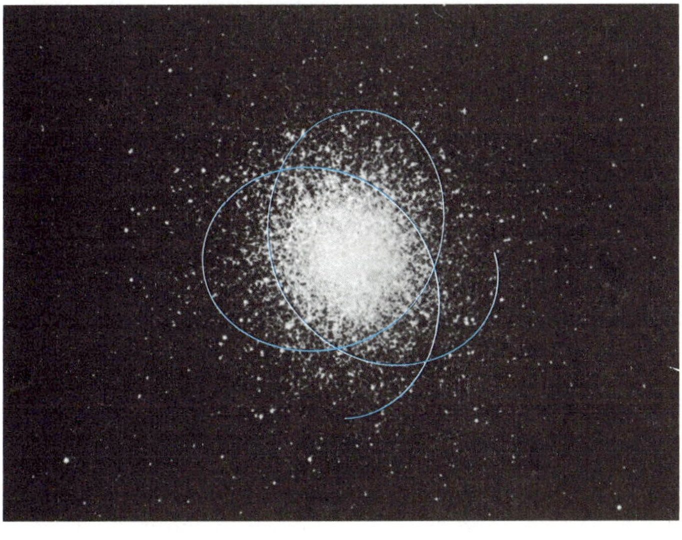

Figure 4.1 MESSIER 13 in the constellation Hercules is the brightest globular cluster in the northern sky; on summer evenings it is overhead and just visible to the unaided eye. It contains some 500,000 stars, and the star density at its center is about 20,000 times that of the solar neighborhood. On the false-color image of the central region (*top*) the brightest colors represent the regions of most intense emissions. The image was made with a charge-coupled device (an electronic, silicon-based light detector) attached to the 200-inch telescope on Palomar Mountain. The area covered by the photograph (*bottom*) is 15 minutes of arc wide, about three times as wide as that in the CCD image. Stars orbit the cluster center with a period that is on the order of a million years. The orbit of a typical star (*blue line*) lies nearly in a plane, but the star does not return to its original position.

dozen globular clusters. He found that they make up an extended system centered behind the brightest star clouds of the Milky Way, in the constellation Sagittarius, where the clusters are commonest. Shapley made what he later described as a "bold and premature assumption": that the globular clusters constitute a kind of "bony frame" whose centroid lies at the center of the entire stellar system. The sun, he argued, actually lies far from the center, toward one edge of the disk.

Shapley had in effect discovered the Milky Way galaxy. The galactic system defined by the spatial arrangement of the globular clusters was much larger than the apparent "local system," which was later shown to be an illusion caused by the murkiness of space. Interstellar dust absorbs starlight, making the stars appear more distant than they really are. When interstellar absorption is neglected, astronomical distances are overestimated; the magnitude of the error increases with the actual distance of the object. As a result fainter, faraway stars appear to be distributed much more sparsely in space than bright, nearby stars, producing the illusion of a systematic falloff in star density in all directions from the earth. It was this illusion that buttressed the heliocentric conception.

Ironically, Shapley himself did not take interstellar absorption into account; but he had the good fortune to be observing clusters well outside the absorbing layer of dust, which is largely confined to the thin, flat disk of the galaxy. Nevertheless, his neglect of absorption led him to overestimate greatly the cluster distances and hence the size of the galactic disk, to which he assigned a radius of 50,000 parsecs. The error was corrected only in 1930, when Robert J. Trumpler of the Lick Observatory showed that interstellar absorption is a general phenomenon. Today the radius of the galactic disk is put at about 15,000 parsecs (see Figure 4.2).

Shapley's "premature assumption" that the centroid of the globular-cluster distribution defines the center of the galaxy is now accepted as fact. A strong source of radio emissions in Sagittarius clearly marks the direction to the center, but an accurate value for its distance from the sun has been elusive. Shapley's basic approach of plotting globular-cluster positions has been applied repeatedly. A comprehensive survey in 1976 by William E. Harris of McMaster University in Ontario yielded a value of 8,500 parsecs. More recently, however, Carlos Frenk of the University of Sussex and Simon D. White of the University of Arizona have argued that globular-cluster distances continue to be overestimated and that the center of the galaxy is actually only about 6,800 parsecs away. The correct figure probably lies somewhere between these two values. Although other methods of finding the center have been used, with similar results, globular clusters still seem to offer the best hope of settling the issue.

The "bony frame" made of globular clusters is also a good tracer of the outline of the Milky Way. The reason has to do with the distinctive stellar content of the clusters, which Shapley discovered when he measured the color and magnitude of individual cluster stars. He noted that the distribution of the stars on a Hertzsprung-Russell diagram (on which the vertical axis represents magnitude, or luminosity, and the horizontal axis represents color) is quite different from that of ordinary stars of the solar neighborhood. For example, the brightest nearby stars are blue, whereas those in globular clusters are red.

STELLAR POPULATIONS

No one made much of this curiosity until 1944. Then Walter Baade of Mount Wilson observed that the brightest stars in the central region of the Andromeda galaxy are also red giants. He thereupon proposed, in a great imaginative leap, that all stars are divided into two fundamentally different "populations." Population I consists of the solar-neighborhood stars and in general of the stars found in the disk of the Milky Way and of other galaxies. Population II stars are scattered throughout an almost spherical "halo" surrounding the disk, although like the disk stars their concentration is greatest toward the center of the galaxy. Globular clusters are simply luminous, readily observable aggregations of halo stars. Thus their distribution not only points the way to the galactic center but also traces the extent of the halo; it is now thought the Milky Way halo may reach as much as 100,000 parsecs from the center.

When Baade conceived the notion of stellar populations, the physical basis of the observed differences in their color-magnitude distributions was not immediately obvious. It began to become clearer after World War II, as the technique of photoelectric photometry came into wide scientific use. Photoelectric studies of star clusters produced much more accurate color-magnitude diagrams. A Hertzsprung-Russell diagram of a cluster, whose stars were presumably all formed at roughly the same time and place, in effect shows the track of stellar evolution: the stars are spread out on the upper part

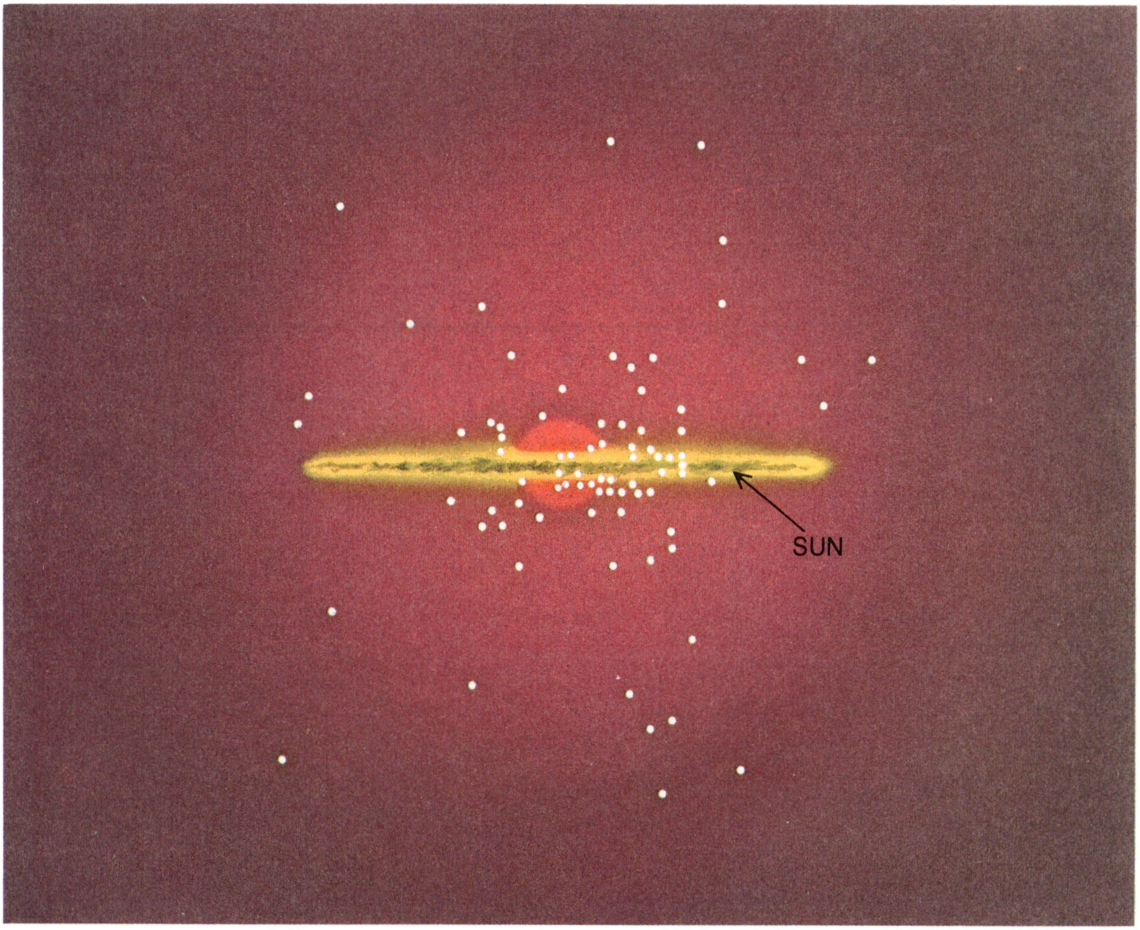

Figure 4.2 SOME 125 GLOBULAR CLUSTERS (*white dots*) of the Milky Way are known; those whose positions are known are shown here. Dust may obscure our view of many more on the far side of the galactic plane from the sun. The cluster distribution defines the center of the galaxy. Globular clusters are aggregations of Population II stars: stars belonging to the galactic halo (*red*), which is thought to have formed as the protogalactic gas cloud collapsed not long after the big bang. The young stars of the spiral-armed disk (*yellow*), where gas is still present and star formation continues, belong to Population I. The radius of the disk is roughly 15,000 parsecs.

of the diagram because the brighter and more massive ones evolve faster. Most of a star's life is spent on the "main sequence" of the diagram, during which time it radiates energy by fusing hydrogen into helium in its core. When the supply of hydrogen in the core is exhausted, the star "turns off" the main sequence, continues to burn hydrogen in a thin shell around the core and evolves into a red giant.

The position of the turnoff point is an index of the cluster's age: the brightest stars turn off first, and as the cluster ages, the turnoff point moves down the main sequence into regions of lower luminosity. In the early 1950's a number of workers estimated the ages of both globular clusters and "open" clusters, which are much less dense aggregations of Population I stars in the galactic disk. These studies revealed the first physical difference between the two stellar populations. Individual open clusters turned out to span a range of ages, but the globular clusters were all older—indeed, they seemed to be the oldest objects in the universe (see Figure 4.3).

Further research showed that age is not the only peculiarity of Population II stars: they are also dif-

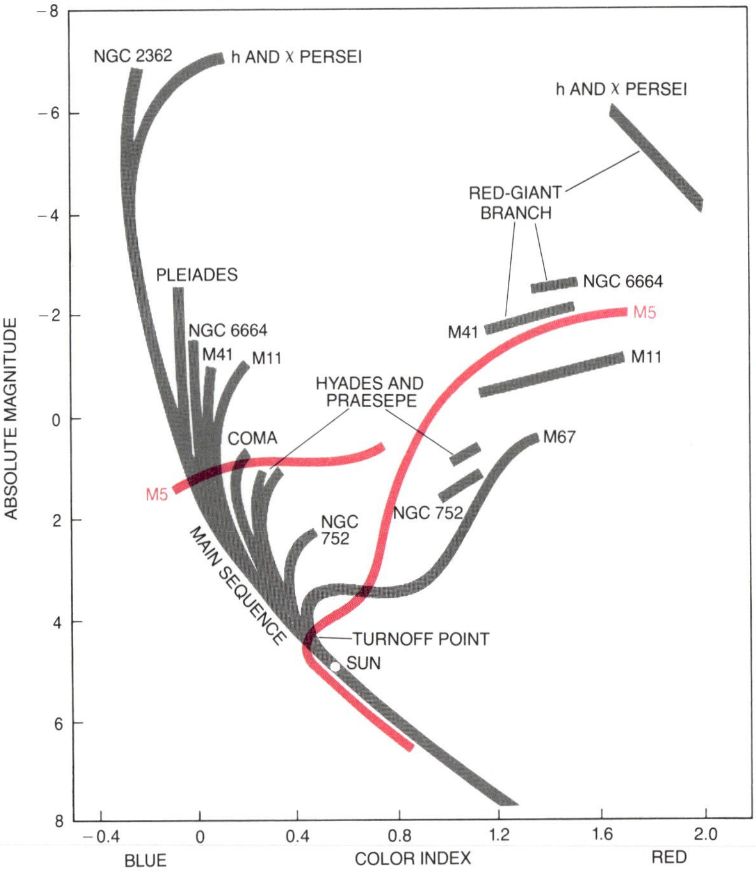

Figure 4.3 HERTZSPRUNG-RUSSELL DIAGRAMS show the distribution of stars in a cluster according to color and magnitude. This schematic composite figure contrasts the H-R diagrams of open clusters (*gray*) in the galactic disk with the diagram of the globular cluster M5 (*color*), which belongs to the halo. Massive, bright stars (those with negative magnitudes) evolve the fastest; they are the first to exhaust the hydrogen in their core, turn off the main sequence and become red giants. As the cluster ages, the turnoff point moves down the magnitude scale. Open clusters (Population I) vary in age. Globular clusters (Population II) are all thought to have been formed at least 13 billion years ago.

ferent in chemical composition. In the mid-1950's Joseph W. Chamberlain, then at the University of Chicago, and Lawrence Aller, then at the University of Michigan, observed that certain dark absorption lines in the spectra of halo stars are weaker than the comparable lines of Population I stars. The weaker lines indicated lower abundances in the halo population of the chemical elements that absorb radiation at those particular frequencies.

The elements in question turned out to be the "heavy" elements: all elements except hydrogen and helium. In calculating the first stellar-evolution tracks for globular clusters, Fred Hoyle, then at the University of Cambridge, and Martin Schwarzschild of Princeton University found that the observed red-giant tracks could be explained only by assuming a deficiency of heavy elements. In astronomical shorthand such elements are called metals, even though the most abundant ones are carbon, oxygen and nitrogen. Hence globular clusters, in addition to being old, were identified as "metal-poor."

Most astronomers now believe globular clusters, and halo stars in general, are metal-poor precisely because they are old. It is generally accepted that the big bang with which the universe began created only hydrogen and helium. The heavy nuclei are thought to have been synthesized at a later time inside stars, where the prevailing temperature and pressure are high enough, and then ejected into space by supernova explosions; the heaviest elements may have been formed during the explosions themselves. According to this scenario, the globular clusters and the halo stars formed early on, while the abundance of heavy elements was still low throughout the universe. The protogalactic gas clouds then collapsed to form the disks of the Milky Way and of other similar galaxies. By that time nucleosynthesis in dying stars of the halo population had raised the heavy-element abundance of the gas to its present level. The Milky Way halo is probably a good indicator of the original size of our galaxy at the time stars began to form; today star formation continues only in the gas clouds of the thin disk.

POPULATION COMPLICATIONS

The foregoing explanation of stellar population differences is appealingly simple, but unfortunately the truth is a bit more confusing. Globular clusters are not all alike. To begin with, although they are all metal-poor, the specific abundance of heavy elements varies from cluster to cluster. This was first noted nearly 30 years ago by William W. Morgan of the Yerkes Observatory, who observed differences among clusters in the strength of their spectral lines. Since then investigators have made quantitative estimates of heavy-element abundances by analyzing in detail the spectra of individual stars. Such analyses have shown that metal concentrations in globular-cluster stars range from about one two-hundredth of the levels observed in the sun (a typical Population I star) to only slightly less than solar values. The precise upper limit is still uncertain, primarily because the spectra of individual cluster stars are quite faint.

Certain stars ordinarily assigned to the halo population even seem to have metal abundances equivalent to those of the sun. The RR Lyrae stars are variable stars (ones whose brightness changes periodically) that are common in globular clusters and throughout the Milky Way halo, and some of them have heavy-element spectral lines as strong as the corresponding solar ones. If there were indeed a continuum of metal abundances in halo stars extending right up to the levels characteristic of disk stars, then the sharp distinction between the two populations would be undermined. Two circumstances, however, suggest that the strong-lined RR Lyraes are not true halo stars. First, no strong-lined RR Lyrae has ever been found in a globular cluster. Second, their orbits around the galactic center are closer to those of disk stars than to those of halo stars. In spite of their striking visual resemblance to their brothers in the globular clusters, the strong-lined RR Lyrae stars may belong to a separate stellar class, perhaps even to a population intermediate between the disk (I) and halo (II) populations.

In any case globular clusters do exhibit a range of metal abundances; might this be evidence of a range of ages? Although it seems very likely that the globular clusters of the Milky Way halo all predate the more numerous stars of the disk, it is not clear just how old they are, and whether the period of their formation spanned an appreciable fraction of the early history of the galaxy. The method of estimating a cluster's age has remained essentially unchanged since it was developed in the 1950's: one looks for the age that, in conjunction with a theoretical model of stellar evolution, best reproduces the observed distribution of cluster stars on a color-magnitude diagram, particularly at the turnoff point. Don A. VandenBerg of the University of Victoria in British Columbia has recently computed an impressively precise set of evolutionary tracks for a number of globular clusters. He has concluded that the clusters are all approximately 16 billion years old, but even these calculations contain an uncertainty of about three billion years (see Figure 4.4).

Some astronomers contend that, whatever the age of the Milky Way globular clusters, they must all be equally old, because dissipation of energy in the star-forming gas cloud would have prevented it from maintaining the spherical shape of the halo for a long period. According to this argument, the spinning cloud would have collapsed quickly into the thin disk, leaving behind the globular clusters and the other halo stars. The range of metal abundances could be attributed to the fact that different globular clusters formed in local regions of the gas, whose chemical composition was not uniform.

Yet there is evidence that globular clusters differ from one another in respects other than their heavy-element content: clusters with the same metal abundances often have noticeably different Hertzsprung-Russell diagrams. For example, the "horizontal branch," which follows the red-giant stage on the evolutionary sequence, may contain blue stars or red stars or both. There must be a

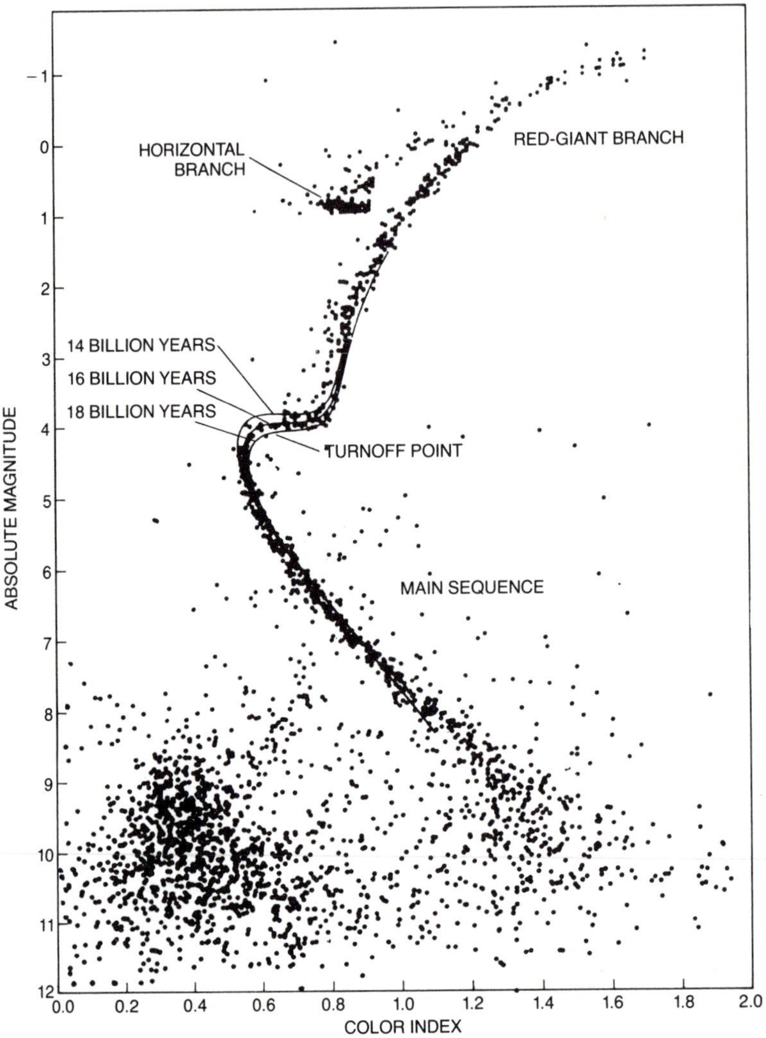

Figure 4.4 AGE OF A GLOBULAR CLUSTER can be estimated by comparing a theoretical model of the evolution of stars in the cluster with their observed distribution on an H-R diagram, as shown here for the giant globular cluster 47 Tucanae. Most of the data were obtained by James E. Hesser of the Dominion Astrophysical Observatory and William E. Harris of McMaster University. The theoretical models (*black lines*) were computed by Don A. VandenBerg of the University of Victoria. The "best fit" at the turnoff point is offered by the model corresponding to a cluster age of 16 billion years. Most of the stars in the lower left corner of the diagram actually belong to the Small Magellanic Cloud, a galaxy behind 47 Tucanae.

second variable parameter that accounts for these differences, and some workers maintain it is age. The differences might also be explained by a variation among clusters in helium content, in the detailed proportions of the heavy elements or in the rate at which the stars spin on their axes. The nature of the second parameter is still a mystery.

COSMOLOGY

Even more of a mystery is how globular clusters acquired any heavy elements at all, given that the big bang is thought to have produced only hydrogen and helium. The observed metal abundances, while quite low compared with those of Population

I stars, are not insignificant. Consequently there must have been an earlier generation of stars inside of which the heavy elements now found in cluster stars were synthesized. So far no traces of a primordial stellar population, in the form of stars older than globular clusters, have ever been detected.

Because the globular clusters of the Milky Way halo are the oldest objects known, their age sets a lower limit on the age of the universe itself. This constraint on cosmological theory is particularly valuable now, at a time when observational cosmology finds itself in a sorry state. Allan R. Sandage of the Mount Wilson and Las Campanas Observatories once described cosmology as "the search for two numbers": the present expansion rate of the universe and the rate at which the expansion is decelerating. (That the universe is in fact expanding is proved by the red shift in the spectra of distant galaxies, which shows that they are receding from our own galaxy at a velocity proportional to their distance.) Neither number is at all well known from observation. The present expansion rate, usually referred to as the Hubble constant, is in principle the easier of the two to calculate, and yet a dispute over its value has divided cosmologists into rival camps. One camp, led by Marc Aaronson of the University of Arizona and Jeremy R. Mould of the California Institute of Technology, puts the Hubble constant between 80 and 100 kilometers per second per million parsecs; Sandage and his colleague Gustav A. Tammann say the correct value is just over half that large, roughly 55 kilometers per second per megaparsec.

The value of the Hubble constant is directly related to the age of the universe, because by extrapolating the expansion rate back into the past one arrives at the time the expansion began. It is virtually certain that the mutual gravitational attraction of galaxies has slowed the expansion, which must have been even faster in the past, and so the age extrapolated from the Hubble constant is an upper limit. The higher of the two proposed values implies that the universe can be no more than 10 to 11 billion years old; it thus conflicts sharply with the best globular-cluster data, which indicate cluster ages of at least 13 billion years.

If Sandage and Tammann's lower value for the Hubble constant is correct, the big bang could have taken place as much as 20 billion years ago. Many cosmologists now believe, however, as a result of recent developments in particle physics, that the rate at which the expansion is decelerating—the second cosmological number—is large. In that case even the lower value for the Hubble constant implies that the universe is only 12 to 13 billion years old. This is still uncomfortably low considering the globular-cluster ages. It is not clear how the conflict will be resolved.

GALAXY FORMATION

Globular clusters do not bear only on the date of the big bang; they may also offer clues to how the galaxies formed. Soon after the primordial fireball the uniform, diffuse mass of hydrogen and helium began to fragment into vast clouds. The size of the clouds must have been determined by a balance between gravity, which tended to pull the gas together, and heat, which tended to disperse it. P. J. E. Peebles and Robert H. Dicke of Princeton have suggested that the pregalactic clouds, which formed in great numbers, are most likely to have been the size of globular clusters. The clouds drifted together under their mutual gravitational attraction. Although most of them coalesced into the larger agglomerations that formed galaxies, some of them escaped collision while remaining gravitationally bound to the larger galactic structures. Such clouds, according to Peebles and Dicke, went on to form the globular clusters of the galactic halo.

If this scenario is correct, it is likely that it will have to be modified to explain the formation of galaxies other than the Milky Way. In addition to the 125 or so globular clusters known in our own galaxy, thousands have been identified in each of a number of other galaxies. Most of these distant clusters are quite faint, and the study of them has begun in earnest only recently. Nevertheless, some differences between galaxies have already emerged. For example, in the Milky Way all globular clusters are old, and the young open clusters of the galactic disk are much poorer in number of stars; the Clouds of Magellan, in contrast, contain young, star-rich aggregations that much resemble globular clusters. It is not known why the Magellanic clouds, our nearest galactic neighbors, should still be making rich clusters when the Milky Way is not.

Nor is it understood why elliptical galaxies seem to have many more globular clusters per unit of mass than spiral galaxies. The observation is of particular significance because it argues against a popular theory of how the ellipticals formed. Alar Toomre of the Massachusetts Institute of Technology and other investigators have proposed that elliptical galaxies are formed when spiral galaxies collide and merge. The strongest evidence against this hypothesis is the higher proportion of clusters in the ellipticals.

Every bit as fascinating as the implications of globular clusters for the structure and formation of galaxies is the structure of the clusters themselves. How does the interaction of thousands of stars produce an overall form of such simplicity and regularity? Each star is held in the cluster by the joint gravitational attraction of all the others; it loops inward and outward in a regular rose-shaped orbit whose period is on the order of a million years. On the average at a given moment half of the stars are moving outward and half are moving inward. Their velocities are just large enough to balance the gravitation that would otherwise draw them into the center. More precisely, there is an exact correspondence between the distribution of stellar velocities and the radial distribution of stars, which determines the cluster's density profile and thereby its gravitational field.

In principle many different pairings of these two quantities are possible, but the structural similarity of most globular clusters suggests that certain velocity and density distributions are favored. The favored distributions arise from the nature of stellar interactions in a cluster. Although the motion of each star is governed almost completely by the rather smooth gravitational field of all its cohorts, on rare occasions two stars pass close enough to each other to affect each other's motion individually. The exchange of energy arising out of such random stellar encounters tends to produce what is called a Maxwellian distribution of velocities, after the Scottish physicist James Clerk Maxwell, who derived a statistical formula to describe the motions of molecules in a gas.

A globular cluster cannot achieve a full Maxwellian distribution, which would include objects of all velocities, because the cluster has a finite escape velocity; stars accelerated to a higher speed by stellar encounters acquire enough energy to escape the cluster's gravitational field. Below this cutoff, however, the distribution of stellar velocities in a cluster closely approximates Maxwell's formula. The velocity distribution in turn determines the radial density profile.

Of course globular clusters are not all structurally identical. Two decades ago I studied many of them

Figure 4.5 RANGE OF PROPERTIES among globular clusters is illustrated by a photograph of two that happen to lie in the same direction. M53 (*upper right*), in the constellation Coma Berenices, is typical of dense, star-rich clusters that are tightly bound. In contrast, NGC 5053 (*lower left*) is a relatively loosely bound cluster containing far fewer stars.

and found that their structural differences could be adequately described by three parameters: the radius of the central core, the outer radius and the number of stars in the cluster. The most important difference among clusters is in the core radius, which is defined as the radius at which the density of stars on the cluster's image has fallen to half of its value at the center. Some clusters have smaller, denser cores than others; these clusters are more tightly bound, and their escape velocities are correspondingly higher (see Figure 4.5). The escape velocity is just the velocity a star must have to reach the outer radius of the cluster. Unlike the boundary of the core, the outer radius is not determined solely by the cluster's gravitational binding energy. Rather, it is primarily a tidal limit defined by the gravitational field of the galaxy, which tends to pull stars out of the cluster.

DYNAMICAL EVOLUTION

Although globular clusters are obviously long-lived, they are not immutable. Slowly but steadily stars "evaporate" from a cluster as they reach the escape velocity. A theory that adequately predicts the resulting evolution of the cluster is basically simple, but many of the details have remained elusive.

The binding energy of a cluster is really an energy deficit: the energy it would take to accelerate all the stars to their escape velocity and tear the cluster apart. To reach escape velocity a star must acquire enough positive kinetic energy to overcome the cluster's negative gravitational energy. Thus the stars that escape are those with the most kinetic energy, whereas the amount of gravitational energy they contribute to the cluster is no more than average. As a result the evaporation of stars increases the amount of binding energy per star remaining in the cluster, and the cluster contracts.

According to current theory, it does not reach a steady state. Instead, energy from the contraction is converted into the kinetic energy of stellar motion, thereby "heating" the core. More stars evaporate, and the core continues to contract and heat without bound until it is infinitely dense. Donald Lynden-Bell of Cambridge, a proponent of the theory, has dubbed this positive feedback phenomenon the "gravothermal catastrophe."

When the theory was first put forward in 1960 by Michel Hénon of the Nice Observatory, there was little observational evidence to support it. Only one globular cluster, M15, showed any sign of the sharp density peak one would expect in a collapsed core. Recently, however, my colleague Stanislav Djorgovski and I have been making more careful observations of a large number of globular clusters. We have observed at least half a dozen displaying central density peaks we believe to be evidence of core collapse (see Figure 4.6). Still, half a dozen is not very many; theories of cluster evolution predict that a much larger fraction of the ancient globular clusters in the Milky Way halo should have collapsed by now. (The various analytic and numerical models also concur in predicting that once core collapse begins it proceeds so rapidly one would be unlikely to observe it in progress.) Why have more central density peaks not been detected?

One possibility is that the predicted time scales for core collapse are too short. A more likely explanation is that some mechanism halts the collapse and even causes the core to reexpand to a normal size.

BINARY STARS

Binary stars—pairs of stars gravitationally bound to each other in a close orbit—could serve as such a mechanism. As long ago as 1959 numerical simulations by Sebastian von Hoerner of the National Radio Astronomy Observatory demonstrated that binaries tend to form in star clusters as a result of chance encounters involving three stars. In later simulations by Sverre Aarseth of Cambridge the contraction of the core was almost invariably stopped by the formation of a massive central binary, whose encounters with other stars gave them "kicks," boosting them into higher orbits.

Aarseth's models were of open clusters containing no more than about 500 stars; Lyman Spitzer, Jr., and Michael Hart of Princeton subsequently showed that three-body encounters are much less likely to produce binaries in globular clusters, which generally contain 100,000 stars or more. In part the reason is that stellar velocities in globular clusters are much greater. As the core of a globular cluster collapses, however, it effectively detaches itself from the surrounding envelope. Thus eventually it might contain sufficiently few stars to allow binaries to form. In addition Andrew Fabian, James Pringle and Martin J. Rees of Cambridge have found that two-star encounters can also result in the formation of a close binary. Either or both of these mechanisms could operate in a small, dense cluster core and produce the binaries needed to halt its collapse.

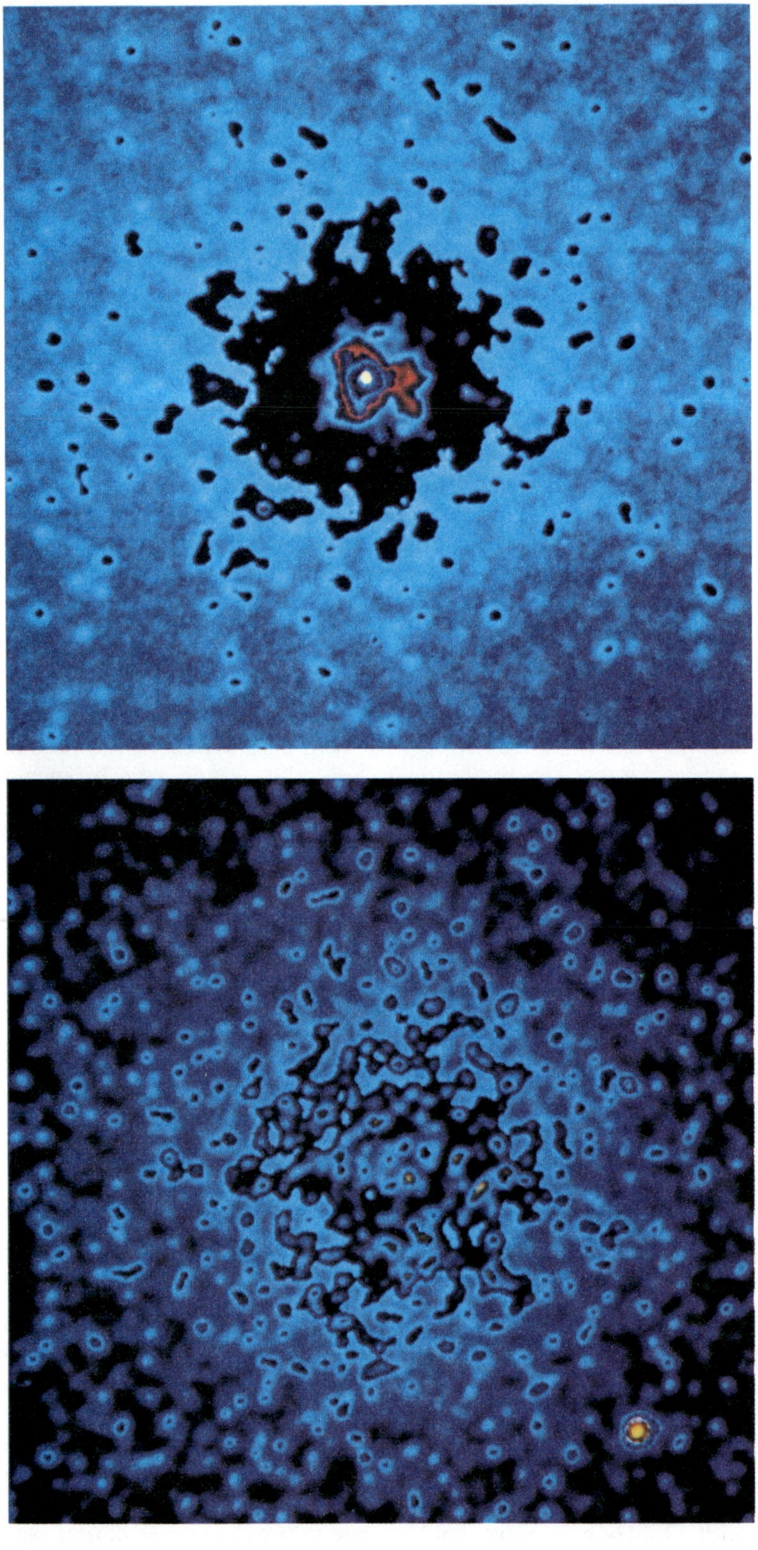

Figure 4.6 CORE COLLAPSE produces a sharp peak in star density at the center of a globular cluster and a corresponding peak in brightness. Such a peak is visible on the computer-generated false-color map of the brightness distribution in the cluster NGC 6624 (*top*). No evidence of core collapse is discernible on the map of 47 Tucanae (*bottom*), although the brightness of that cluster also increases toward its center.

By the time the core is small enough it would probably be almost indistinguishable, at current levels of resolution, from a completely collapsed core; it would probably exhibit the kind of central density peak Djorgovski and I have observed. To explain why such peaks have not been detected in more clusters, one must further postulate that the energy contributed by binaries to the stars in a contracted core is enough to reexpand the core. The idea is attractive, but its plausibility has not been fully established.

Unfortunately the centers of globular clusters are too dense to allow any hope of observing binary stars directly with ground-based telescopes. The high-energy X rays emanating from several clusters, however, may be indirect evidence of double stars. At one time it was popular to suggest the X rays come from material crashing through the tremendous gravitational field of a black hole. Yet one would expect to find such a massive object precisely at the center of a cluster, and recent studies by Jonathan E. Grindlay of Harvard University and his co-workers have shown that the X-ray sources in globular clusters are slightly off-center. It now seems most likely that the sources are close binary systems in which material is sucked from a distended star by the strong field of a neutron star or of a white dwarf.

Nearly all the X-ray sources in globular clusters are situated in dense core regions, and so it is natural to suggest they are the binary stars responsible for halting core collapse. Actually there is no reason to assume that the binary stars stabilizing the core would have to be X-ray binaries; indeed, several of the collapsed-core clusters we have found do not have X-ray sources. Conversely, several of the known X-ray clusters lack collapsed cores. A dense core may simply favor the formation of binaries in general because it promotes stellar encounters.

The crowded centers of globular clusters may yield many of their secrets to the Space Telescope. Observations from the ground are limited in their resolution primarily by the unsteadiness of the earth's atmosphere. The 2.4-meter-diameter orbiting telescope will have a resolving power about 20 times as great as that of the best ground-based instruments.

Today, in the pre-Space-Telescope era, the most rapid observational advances are occurring in the study of the motions of individual cluster stars. With new digital spectrographs it is now possible to measure a star's motion along the line of sight by means of the Doppler shift in its spectrum. This technique is more accurate than the cruder method of measuring transverse velocities from the tiny displacements of stars on photographs taken decades apart. Doppler measurements are adding a new dimension to knowledge of cluster structure.

Like all endeavors in astronomy, the study of globular clusters has benefited tremendously from such technological improvements. Yet globular-cluster investigations in particular have profited from their position at the intersection of different avenues of research, just as they have been essential to many fundamental developments in astronomy. It is this unique position, more than any particular technology, that allows the student of globular clusters to be sanguine about the future. New insights could come from just about anywhere.

Molecular Clouds, Star Formation and Galactic Structure

5

Radio observations show that the giant clouds of molecules where stars are born are distributed in various ways in spiral galaxies, perhaps accounting for the variation in their optical appearance.

• • •

Nick Scoville and Judith S. Young

April, 1984

Most of the stars visible in the night sky were formed more than a billion years ago. Star formation is a continuing process, however, and much of what is fascinating about the universe has to do in one way or another with the comparatively small number of much younger stars. The most massive, short-lived members of each stellar generation are of particular interest. In youth they energize the fluorescent nebulas that stud the spiral arms of galaxies such as our own. In death they explode spectacularly as supernovas, replenishing the interstellar environment with a mixture of gases, including an enriched fraction of heavy elements. It is from these ashes that future generations of stars will arise. In recent years it has become clear, mainly from observations made by means of radio telescopes, that the springs of this rejuvenation are giant molecular clouds, measuring more than 100 light-years across and encompassing a mass of gaseous material up to a million times the mass of the sun. Inside these immense cocoons the metamorphosis of stars takes place in cold and dusty darkness.

Although the giant gas clouds are known to be fertile sites for the formation of stars, it is significant that they have not been completely assimilated into stars. Today, more than 10 billion years after the birth of our galaxy, one can still see many young stars emerging from the clouds in which they were formed. Isolated from the galactic environment, the clouds would collapse under their own weight, transforming their diffuse gas into stars in less than 10 million years. If star formation were inevitable, requiring merely that an adequate mass of material be accumulated into a cloud, the supply of interstellar gas needed to form the next generation of stars would have dwindled long ago to insignificance. Thus a subtle interplay of the clouds and their galactic environment must effectively regulate the formation of stars. One of the most exciting episodes of contemporary astrophysics has been the observational and theoretical work done in an attempt to define the role of the giant molecular clouds in galactic evolution and to establish the links between star formation and the large-scale structure of galaxies.

Figure 5.1 M101 SPIRAL GALAXY is estimated to be approximately 20 million light-years away and has a span across the line of sight of about 300,000 light-years. The spiral arms are illuminated by bright ionized-hydrogen, or H II, regions, which are heated to fluorescence by massive young stars embedded in them. The dark filaments in the spiral arms are dust lanes. Radio observations of carbon monoxide (CO) molecules in the vicinity of M101 reveal that the star-forming molecular clouds there are concentrated toward the center of the galaxy, extending outward to a radius of roughly 40,000 light-years.

The proportion of stars that are young can vary greatly both from galaxy to galaxy and from place to place in an individual galaxy. In spiral galaxies such as ours the arms appear fairly bright in photographs made in visible light, owing to the concentration of massive, young stars along their length. Although such stars have a comparatively brief lifetime (less than 10 million years), their rate of radiation can be a million times that of the sun. Hence their birthplaces will be brightly illuminated for a few million years. One of the most remarkable features of spiral galaxies, first recognized 40 years ago by Walter Baade of the Mount Wilson Observatory, is the apparent correlation in the positions of massive stars such that a grand spiral pattern is perceived. The bright arms can often be traced over a complete turn, spanning a distance of perhaps 200,000 light-years (see Figure 5.1).

How is it that the formation of massive stars can be correlated over the entire galactic disk, a range well beyond the physical effects of one cloud on another or the sphere of influence of an individual star? There are now two schools of thought on the question. It was first proposed by Chia-Chiao Lin and Frank H. Shu of the Massachusetts Institute of Technology that these large-scale patterns are density waves generated by the collective gravitational interactions of billions of stars in the galactic system or by the tidal pull of a nearby galaxy. More recently Philip E. Seiden and Humberto C. Gerola of the IBM Thomas J. Watson Research Center have suggested as an alternative that star formation may spread across the face of a galaxy like a forest fire, with the formation of massive stars at one location setting off the formation of other stars in adjacent clouds. (A third possibility is that the clouds exist and form stars throughout the galactic disk, but that near the spiral arms their properties change and they preferentially form the more massive stars there.)

For years astronomers have been able to identify gas clouds near the solar system as sites of active star formation. Viewing the actual birth process, however, was not possible until recently. A small admixture of dust in the clouds, constituting about 1 percent of the mass, effectively absorbs the visible and ultraviolet radiation from embedded young stars in all but the most tenuous clouds. The microscopic dust particles are composed of carbon (in the form of graphite), silicates and other compounds similar to some terrestrial and lunar rocks. Although the dust in interstellar space is quite sparse, it is readily apparent when one gazes at the Milky Way on a dark night. One then sees the disk of our galaxy not as a single, smooth band of stars across the sky but as two bands, with a dark void between them. The rift appears because the light from more distant stars in the galaxy is absorbed by clouds of gas and dust lying in the foreground along the line of sight.

The layer of galactic dust seen silhouetted against the background stars has a thickness of 300 light-years. Because of the dust, it has been impossible to observe visually the very youngest stars until they either drift away from the enshrouding cloud or release enough energy to disperse it. As the stars age they interact gravitationally with the massive interstellar clouds, gradually increasing the distance by which they are flung outward from the galactic disk before they fall back. As a result, although most of the stars were probably formed out of gas clouds in the thin disk, the thickness of the older stellar layers has increased to about 1,000 light-years.

One of the best-known stellar nurseries in our galaxy is the Great Nebula in Orion (see Figure 5.2). The luminous nebula is easy to see with binoculars in the middle of the dagger below Orion's Belt. Astronomers now recognize that the bright emission from the nebula, first studied more than a century ago, is a manifestation of the final phase in the process of star formation. Near the center of the nebula's emission is the Trapezium star cluster, which includes several massive young stars. The most massive of them radiates energy at a rate 100,000 times that of the sun. A substantial fraction of the radiation is in the ultraviolet region of the

Figure 5.2 VIEW OF THE ORION NEBULA at top left was made with the four-meter telescope on Kitt Peak. The false-color image at the top right, covering the same region, was made from a record of the radio waves emitted by carbon monoxide molecules in the vicinity of the nebula. The photo at bottom left, made with the three-meter telescope at the Lick Observatory, is the central part of the nebula outlined by the square in the two top views. This photo reveals the Trapezium cluster that is responsible for the nebula's fluorescence. The corresponding image of the central region at bottom right was made with the NASA infrared telescope at Mauna Kea in Hawaii. The bright spots represent the youngest stars in the nebula, still embedded in the central dust cloud and hence invisible. (Infrared data were obtained by Gareth Wynn-Williams, Eric E. Becklin, Reinhart Genzel and Dennis Downes.)

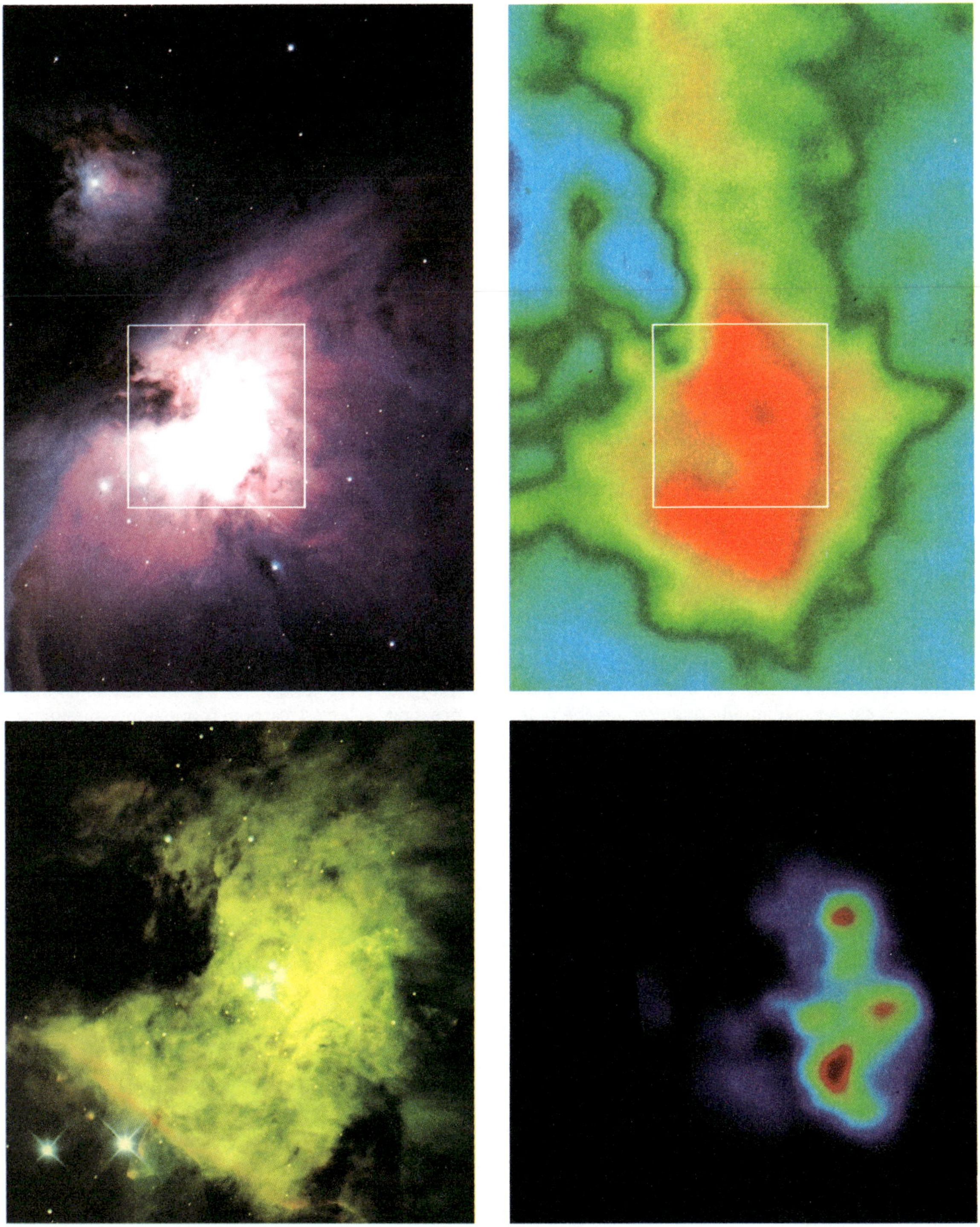

spectrum, and hence it is sufficiently energetic to strip electrons from atomic hydrogen, thereby ionizing the hydrogen in the surrounding gas. The bright visible emission, characteristically red in color, is actually fluorescent light at a wavelength of 6,563 angstrom units emitted by the ionized hydrogen when it recombines, capturing a free electron.

For the entire nebula this cycle of ionization and recombination is repeated approximately 10^{50} times per second. Thus in spite of the fact that the region is some 1,700 light-years from the solar system, it is observed to glow brilliantly. In the nebular gas the ionization of hydrogen atoms proceeds about 1,000 times faster than the reverse process: the recombination of electrons and protons to form hydrogen atoms. Hence a state of ionization equilibrium exists in the nebula at a concentration of electrons and protons 1,000 times that of neutral (un-ionized) atomic hydrogen. Regions such as the Orion nebula, where the bulk of gas is ionized, are designated H II regions; they are the bright nebulas that delineate the spiral arms in galaxies.

As one of the nearest H II regions, the Orion nebula has become a Rosetta stone for studies of star formation and associated physical processes. Scanning the breadth of the visible nebula, one sees both stationary clumps and high-speed streams of plasma (ionized gas). Moving farther out from the Trapezium cluster, eventually one reaches the point where all the ultraviolet photons emitted by the four Trapezium stars have been exhausted in ionizing the gas. Beyond this radius the hydrogen can no longer be kept ionized. Until recently the ionized-hydrogen region was the only part of the nebula that could be observed. Strong obscuration by the grains of dust mixed with the neutral gas made it impossible to see inside the cloud.

Now developments in infrared astronomy and millimeter-wave radio astronomy have shown that the earlier observations at visible wavelengths were revealing little more than the tip of the iceberg. Both the infrared radiation and the millimeter-wave radiation are much less attenuated by the dust. The former makes it possible to detect stars deeply embedded in the dust; the latter provides a means of observing the molecular gas. The infrared data are in a sense a negative of the optical picture, since the dust that absorbs the visual light will subsequently reradiate the excess energy at infrared wavelengths. Thus in the infrared part of the spectrum one sees bright emission associated with luminous stars embedded in the dust cloud; one rarely sees the star itself but rather the reradiated energy from the heated dust nearby. Infrared observations of the Orion nebula have revealed two clusters of young stars deep in the neutral cloud. The fact that neither of the new clusters was even faintly apparent in visible-light photographs of the region is remarkable; inasmuch as one of the clusters is apparently emitting energy in the form of infrared radiation at a rate nearly identical with that of the Trapezium cluster, or 100,000 times the rate of the sun.

Inside the cloud behind the Orion nebula the gas is sufficiently dense and the temperature sufficiently low for most of the atoms to have become bound as molecules. Here, where the average temperature is lower than 100 degrees Kelvin (degrees Celsius above absolute zero), the most abundant constituent is molecular hydrogen (H_2). There are also numerous trace molecules such as carbon monoxide (CO), cyanogen (CN) and ammonia (NH_3). At the temperatures generally prevailing in the clouds the molecular hydrogen is not directly observable. Studies of the cool gas have relied on the trace molecules; unlike H_2, they can emit and absorb radiation at short radio wavelengths and in the far-infrared part of the spectrum. An important phase in the understanding of star-forming clouds was ushered in 15 years ago when Robert W. Wilson, Keith B. Jefferts and Arno A. Penzias of Bell Laboratories first detected carbon monoxide emission in the Orion cloud. Since then about 60 molecules ranging in complexity up to cyanopentacetylene ($HC_{11}N$) have been identified in such regions. The list includes formic acid (HCOOH), formaldehyde (H_2CO) and ethanol (C_2H_6O). Because the more complex molecules are not as abundant as CO, they can usually be detected only in the compact core of the clouds, where the gas is densest.

Carbon monoxide remains the best tracer of molecular gas over wide areas. In the Milky Way the mean density of interstellar matter is approximately one atom per cubic centimeter. The molecular clouds, which are the comparatively dense parts of the interstellar medium, have typical densities of from several hundred to several thousand molecules per cubic centimeter—still only a billionth of a billionth the density of the earth's atmosphere at sea level. Within the clouds carbon monoxide accounts for approximately one molecule for every 10,000 hydrogen molecules.

The binding of a molecule such as carbon monoxide results from the fact that the outermost electrons of the carbon and oxygen atoms are shared; each of the electrons spends some time near the other atom. Since the sharing is not exactly equal, there is a small net positive charge at one end of the molecule and a similar negative charge at the other end. It is the attraction between the opposite charges that keeps the atoms bound together.

The radiation detected from molecules in the interstellar clouds arises from changes in the rotation of the molecule as a whole; the permissible rotations are quantized rather than continuous. When the carbon monoxide molecule changes from a higher rate of rotation to a lower one, it radiates a photon, or quantum of electromagnetic radiation, with an energy equal to the reduction in the rotational energy. A transition from the first excited energy state of carbon monoxide to the ground state gives rise to a photon with a wavelength of 2.6 millimeters, corresponding to a radio frequency of 115,000 megahertz. Since each molecular species has a slightly different structure, each will radiate at a unique set of frequencies. Their spectral fingerprints are distinct and identifiable in the comparatively short millimeter-wave band.

Opening up this spectral band to astronomical observation has called for large technical investments at the National Radio Astronomy Observatory, Bell Laboratories and several universities where sensitive radio receivers and telescopes with sufficiently accurate reflecting surfaces have been built. Although the technology at short radio wavelengths was initially very taxing, there is a side benefit: even a modest-size telescope provides excellent angular resolution. A 14-meter telescope observing the 2.6-millimeter carbon monoxide line has a resolution of 50 seconds of arc. To obtain equivalent resolution observing the standard 21-centimeter line of atomic hydrogen would require a telescope nearly a mile in diameter.

From the carbon monoxide emission observed in a molecular cloud one can infer not only the density and the temperature of the molecules but also their motions. Motions of the gas along the line of sight are detected by measuring the Doppler shift in the frequency of the emission in a particular parcel of gas away from the frequency of the transition as it is measured in laboratories on the earth. The density and the temperature are inferred less directly. In the absence of external effects the molecule will radiate and decay into the state with the lowest rotational energy, and it will remain there until the environment supplies enough energy to reexcite it. The most important agency for this excitation is collisions with molecular hydrogen. Since the frequency of collision depends on the density of the molecules, the brightness of the molecular emission in a cloud serves as a measure of the density of molecular hydrogen. With carbon monoxide the radiative decays are sufficiently slow compared with typical collision times for the distribution of CO among the various rotational states to have a characteristic thermal distribution in all clouds except those of the lowest density. Hence in the denser clouds the brightness of the CO emission yields an estimate of the temperature in the molecular hydrogen. In such regions one must rely on other molecules, such as HCN (hydrogen cyanide) or CS (carbon monosulfide), which have a faster radiative decay, in order to measure the density of the gas.

The extent of the molecular cloud in Orion can be traced in carbon monoxide emission more than three degrees to the south and two degrees to the north of the bright optical H II region. This angular measurement corresponds to a linear distance of almost 100 light-years in the long dimension. Thus the linear extent of the molecular cloud is about 50 times larger than that of the optically bright region. The quantity of molecular gas in this space is now estimated to be 200,000 times the mass of the sun, or about 1,000 times more than the total mass of the stars visible in the Trapezium cluster.

In the vicinity of the brightest infrared source in the nebula the situation is not at all what one would expect for the gradual collapse of a cold cloud in the process of forming stars. Within about five light-years of the embedded infrared sources their effects are clearly seen. Their radiation heats the surrounding dust, which in turn heats the molecular hydrogen. The carbon monoxide observations show a temperature gradient running from 20 degrees K. at a radius of five light-years to approximately 100 degrees at a radius of .1 light-year. Within .1 light-year an abrupt change appears: gas is moving supersonically (at speeds of up to Mach 100), and a small fraction of the molecular hydrogen is heated to more than 2,000 degrees. Although the cause of this phenomenon is not understood in any detail, it is generally believed that the motions are generated by an energetic young star shredding the last remnants of its natal cocoon. The high temperatures would be at shock fronts where the supersonic gas, thrown outward by the young star, collides with the surrounding cloud. By occasionally stirring up and

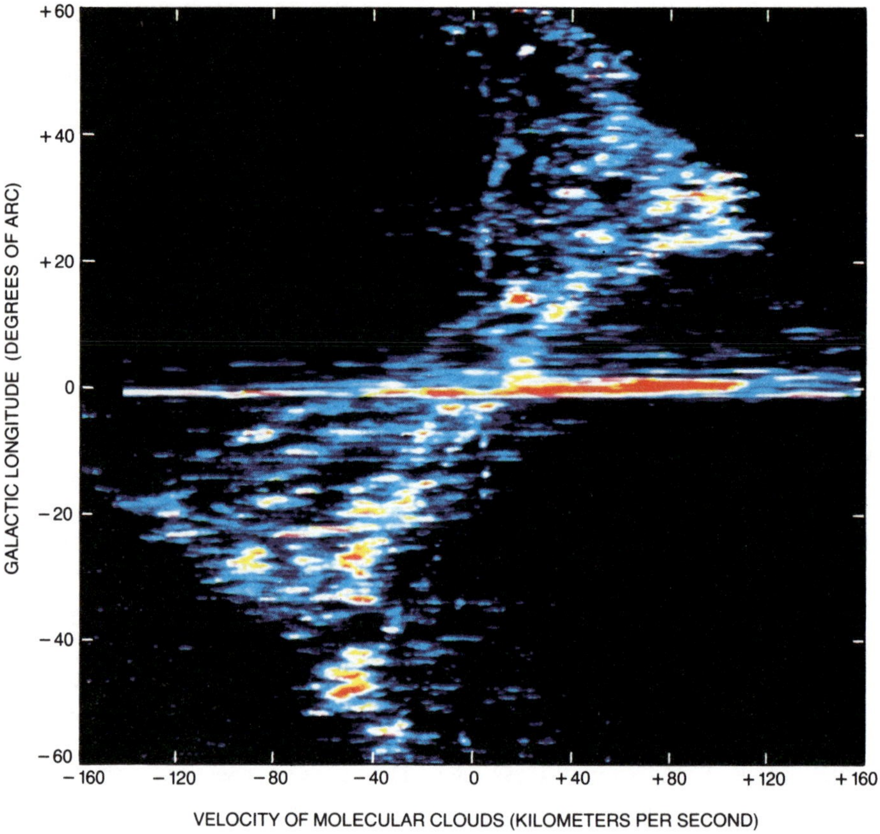

Figure 5.3 MOLECULAR CLOUDS were surveyed by means of the 2.6-millimeter radio waves emitted by carbon monoxide molecules in the inner disk of the galaxy. The individual gas clouds (*colored streaks*) are estimated to be between 50 and 200 light-years in diameter. The vertical scale gives the galactic longitude in degrees of arc; zero corresponds to the direction of the center of the galaxy. The horizontal scale measures the velocity of the individual molecular clouds, on the basis of the Doppler shift in the frequency of the radio waves they emit from a standard frequency for carbon monoxide emission measured in the laboratory. The measurements serve as an indicator of the location of the clouds in the galaxy. (Survey by Daniel Clemens, David B. Sanders, Nick Scoville, Philip M. Solomon, Richard N. Manchester, Brian Robinson, John Whiteoak and William H. McCutcheon.)

disrupting the surrounding cloud such flows may account for the generally low rate of star formation in the clouds. Perhaps the birth of sufficiently energetic stars can effectively limit subsequent star formation in the same area.

The Orion cloud is just one of many such regions in our galaxy. To determine the quantity of molecular gas in the interior of the galaxy one of us (Scoville), working in collaboration with Philip M. Solomon of the State University of New York at Stony Brook, conducted in 1975, the first sampling of carbon monoxide emission from the galactic disk, using the 11-meter telescope of the National Radio Astronomy Observatory. The results bore little similarity to earlier pictures of the more tenuous atomic-hydrogen clouds, designated H I regions. The molecular clouds were found to be extremely plentiful in the central 500 to 1,000 light-years out from the galactic nucleus, but their number fell off at a larger radii. Most surprising was the discovery that the density of the molecular gas rose again to a second peak at a radius about midway from the sun to the galactic center. This ring of molecular gas, with a peak at about 15,000 light-years from the center of the galaxy, also appeared in later, more complete surveys of carbon monoxide emission by radio astronomers at other observatories in the U.S. (see Figure 5.4).

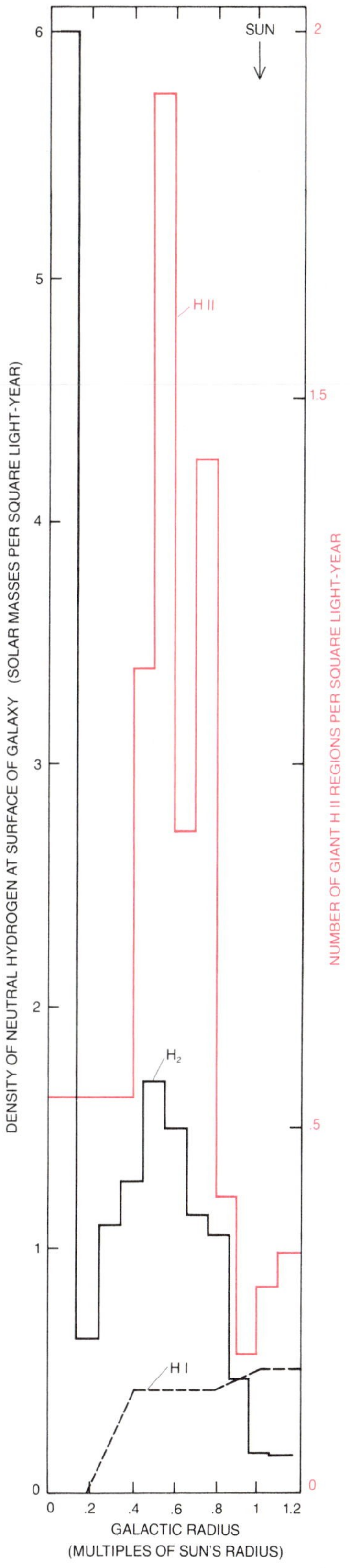

Figure 5.4 EVIDENCE OF A RING of star-forming material in our galaxy is presented in this chart. The line in color shows the distribution of giant ionized-hydrogen (H II) regions, a reliable indicator of where massive stars are forming the galaxy. The black line traces the distribution of molecular hydrogen (H_2). Both distributions exhibit a strong peak at approximately half the distance of the sun from the center of the galaxy. In contrast, the distribution of neutral (un-ionized) atomic hydrogen (H I), indicated by the broken black line, remains fairly constant, rising somewhat beyond the sun's orbit, where there are few giant H II regions.

Since all the early galactic data were collected with radio telescopes in the Northern Hemisphere, a question lingered about the extent to which the southern galactic plane shows a similar structure. Within the past year a group of radio astronomers led by Brian Robinson of the Commonwealth Scientific and Industrial Research Organization (CSIRO) in Australia has completed a study of the emission from carbon monoxide in the Southern Hemisphere. The amounts of molecular gas seen in equivalent areas on both the north and the south side of the galaxy agree to within 20 percent. The southern distribution is similar to the northern one in exhibiting a ring-shaped peak midway between the sun and the galactic center, but the detailed shape of the southern part of the ring is somewhat different: the peak is 30 percent lower in density and the width is correspondingly greater.

The total mass of molecular gas estimated from the carbon monoxide emission in the interior of our galaxy is between one billion and three billion solar masses, which is equivalent to about 15 percent of the total stellar mass in the same region. This amount of gas greatly exceeds the mass of interstellar atomic gas in the region and in fact is comparable to the amount of atomic hydrogen over the entire galaxy out to twice the sun's radius. It should be remembered that until a decade ago this major component in our galaxy was totally unobserved on a galactic scale. It is significant not only for its comparative abundance but also for the fact that the molecular-hydrogen clouds, not the atomic-hydrogen clouds, are the medium from which new stars arise.

Most remarkable are the properties of the molecular clouds. Far from being abnormally large, the Orion cloud is at the small end of the range of the giant molecular clouds in the galaxy. In 1981 David B. Sanders, then a graduate student at Stony Brook, measured more than 300 of the clouds in the galactic ring. He found that most of the gas was contained in clouds with a mean diameter of about 100 light-years. Although the clouds are extremely tenuous (about 300 hydrogen molecules per cubic centimeter), their volume is so great that their total mass amounts to between 100,000 and several million times that of the sun. Indeed, the giant molecular clouds are now thought to be the most massive objects in the galaxy. The number of clouds larger than 50 light-years in diameter is probably about 5,000.

The relationship of the giant molecular clouds to the other components of the galaxy—the massive young stars, the more diffuse gas and the older stars—can provide an important key to understanding the vast galactic machine. Are giant molecular clouds assembled by the collision of numerous smaller molecular clouds or by the compression of diffuse atomic hydrogen? Is there always a molecular cloud near the young stars, and if there is, are the birthplaces of these stars deep within the cloud or near the surface? Stars might form near the surface if the collapse of the clouds is triggered by external factors such as the collision of one cloud with another.

The locations of massive young stars such as those in the Trapezium cluster can be pinpointed throughout the galaxy by the low-frequency radio emission generated in the ionized gas surrounding such stars. As might be expected, there is an excellent correlation between the H II regions and the clouds: virtually every known optical H II region or radio H II region equivalent in size to the Orion nebula has a molecular cloud near it. The largest and hottest clouds tend to be associated with these H II regions. Inasmuch as the spiral arms of galaxies beyond our own are best delineated by the H II regions, the correlation seems to suggest that most of the giant clouds are in the arms of our galaxy. As it happens, there are many more giant molecular clouds in our galaxy than there are known giant H II regions (5,000 as opposed to about 200), and many of the cooler, smaller clouds are clearly not near any H II regions. Therefore one must be cautious in interpreting carbon monoxide observations made at low sensitivity or poor angular resolution. Such data would tend to resemble spiral arms because of an observational bias toward the largest, hottest clouds, even if clouds of all sizes and temperatures were fairly uniformly distributed.

A recent model proposed by John Y. Kwan of Bell Laboratories and the University of Massachusetts at Amherst implies that the existence of particularly large clouds in the arms could be explained if smaller clouds from the zones between the arms collide and coalesce, forming a few superclouds, or cloud clusters, when they reach the arms. If the frequency of cloud collisions is higher in the arms, this could account for the higher rate of massive-star formation there. When clouds collide, it is expected that the compression of the gas at the interface would be followed by the collapse of the cloud fragment into one or more protostars. Such compressed regions would be favorable to star formation because the self-gravity of the cloud fragment would be increased as a result of the higher density.

Once the massive stars are formed their high luminosities heat the surrounding dust cloud. Hence the correlation of hot clouds with the locations of the H II regions is quite understandable. After the superclouds leave the arms they may break up into smaller units, perhaps as a result of the disruptive forces of the hot H II regions.

The observation of widespread molecular gas in our galaxy raises basic questions that can be addressed only by looking at other galaxies. For example, is the ring of molecular clouds at a radius of approximately 15,000 light-years from the galactic center a common feature of other galaxies? Is it a mark of the way our galaxy was originally shaped or has our galaxy simply evolved to this form in the course of aging? Is the comparative abundance of molecular gas seen in the interior of our galaxy a general characteristic of most spiral galaxies? Do the quantity and the distribution of molecules in a galaxy depend on the form of the galaxy? Finally, how is the total luminosity of a given galaxy dependent on the quantity and the distribution of molecular clouds? One might expect some relation between the two if a large fraction of the galaxy's energy is generated by young stars that form within the clouds.

It is now well established from optical studies that the stellar properties of galaxies follow predictable patterns. In his pioneering work on galaxies Baade observed that the stars can be divided into two classes. One class, whose members were young, blue stars, he called Population I; the other, whose members include old, red stars, he called Population II.

In the elliptical and lenticular galaxies the stars are almost all older than five billion years. Except in a few instances there is little evidence of either the young, Population I stars or substantial interstellar gas to form future generations of stars; in such galaxies the supply of gas either was exhausted long ago or has been swept out into the surrounding intergalactic space. These galaxies contain primarily Population II stars.

In spiral galaxies such as ours both populations are present. The old, Population II stars occupy a spheroidal volume resembling an elliptical galaxy, whereas the young, Population I stars are found almost exclusively in the thin disk along with nearly all the remaining interstellar matter. Both the spheroidal distribution of old stars and the younger disk have a common center. Among spiral galaxies the relative blend of the old and the young components can vary greatly, ranging from the early-type spirals (designated *Sa*), which have a large nuclear bulge and tightly wound spiral arms, to the late-type spirals (*Sd*), which have an almost insignificant visible nucleus and a very open, patchy spiral pattern (see Figure 5.5). Our galaxy is thought to be an *Sbc* type with intermediate characteristics. Within each morphological class there is a wide range of mass and luminosity, typically covering a factor of between 10 and 100 in total energy output.

Early studies of the radiation emitted by molecules in other galaxies were conducted by Lee J. Rickard of Howard University and Patrick Palmer of the University of Chicago. These astronomers concentrated initially on investigating the galaxies that have abnormally strong infrared emission. In

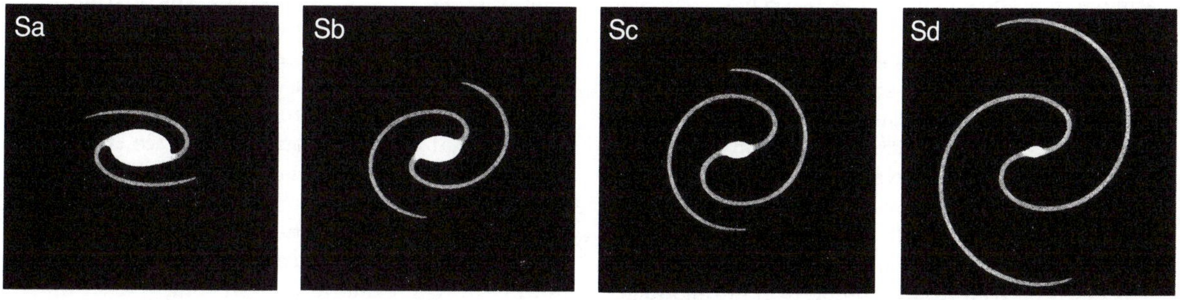

Figure 5.5 SPIRAL GALAXIES range from type *Sa* galaxies, which have a tight arm structure and a large central bulge, to type *Sd* galaxies, which have an open arm structure and a fairly small central bulge. Our galaxy is believed to have an intermediate, type-*Sbc* structure.

these galaxies, it was believed, a burst of star formation was fueled by a plentiful supply of molecular gas. Recently we have undertaken a comprehensive program to map the carbon monoxide emission in more normal spiral galaxies with the goal of elucidating the relation of the content and distribution of molecular gas to the morphology and luminosity of the galaxy. Our observations were made with the 14-meter radio telescope operated by the Five College Radio Astronomy Observatory in Massachusetts; it is the largest such instrument in the U.S. and provides a high angular resolution, enabling us to observe fine details in the external galaxies. Of the 80 galaxies we have studied, almost 40 have shown detectable carbon monoxide emission, and 20 of them have been partially mapped. Most of these galaxies are classified as normal spirals, although a few are irregular.

Because of the great distances of the external galaxies, it is not possible to observe individual molecular clouds in them. The resolution of the 14-meter telescope (50 seconds of arc) does, however, enable us to look at the composite emission from regions typically 5,000 light-years across, encompassing many molecular clouds. The carbon monoxide observations of external galaxies therefore yield a determination of the global distribution of molecular clouds, not their individual properties.

One of the spiral galaxies studied most closely by several groups of radio astronomers is M51, the Whirlpool galaxy. Here the carbon monoxide emission is detected over the entire visible disk; as in many galaxies the greatest concentration of molecular gas is found near the center. Most surprising, however, is the fact that there is a fairly smooth, systematic decrease in concentration from the center out to where the emission becomes undetectable. In other words, there is no evidence of either a ring or armlike concentrations in the molecular gas. The absence of such concentrations may be attributable in part to insufficient resolution: the arms are quite narrow and have little space between them. The absence of a ring, however, is clearly significant: if there were a feature similar to the one in our galaxy, it would be easily observable.

An important clue to understanding how the rate of star formation varies across the disks of galaxies is derived from a comparison of the molecular distribution in M51 with the distribution of luminosity, particularly the luminosity of the youngest stars. With the aid of a telescope carried on the National Aeronautics and Space Administration's C-141 airborne observatory, James Smith of the Yerkes Observatory has recently made a complete map of M51 in the far-infrared part of the spectrum at wavelengths between 80 and 200 micrometers (see Figure 5.6). This radiation is contributed partially by sources similar to the bright infrared sources behind the Orion nebula, which are presumably young star clusters formed in the past 10 million years and still shrouded in dust. The total luminosity measured by Smith in the far-infrared band is 30 billion solar luminosities within the region of the optical disk out to a galactic radius similar to that of the sun in our own galaxy.

Of fundamental importance is the finding that there is virtually a one-to-one proportionality between the far-infrared luminosity and the carbon monoxide emission at different points in M51. Both fall off smoothly with distance from the center of the galaxy, and the dependence on radius is nearly identical. If the rate of star formation is indicated by the energy output in the infrared band, and if the supply of gas capable of forming stars is indicated by the carbon monoxide emission, then one might conclude that the rate of star formation depends solely on the abundance of the molecular clouds, not on their location in the galaxy.

At first this conclusion seems surprising, because it is expected that the external forces promoting the collapse of the clouds to form stars might depend strongly on distance from the center of the galaxy. Perhaps a natural explanation of the simple correlation between the rate of star formation and the mass of molecular matter is to be found in the nature of the clouds. If the clouds in M51 are primarily giant clouds like those in our galaxy, it is difficult to see how external phenomena such as the expanding shells of supernovas and H II regions could penetrate very far inside the clouds. The inertia of a cloud with a mass a million times that of the sun is simply too great for a significant fraction of the cloud to be affected. In a sense the clouds are already pregnant with star formation, and an external stimulus that penetrates only the surface layers can do little to alter the total rate of stellar birth for the entire cloud.

The proportionality between the rate of star formation and the abundance of molecular gas found in M51 now appears to be a general rule in those late-type, high-luminosity spiral galaxies where the amount of molecular gas exceeds that of atomic gas. For most galaxies complete far-infrared data do not

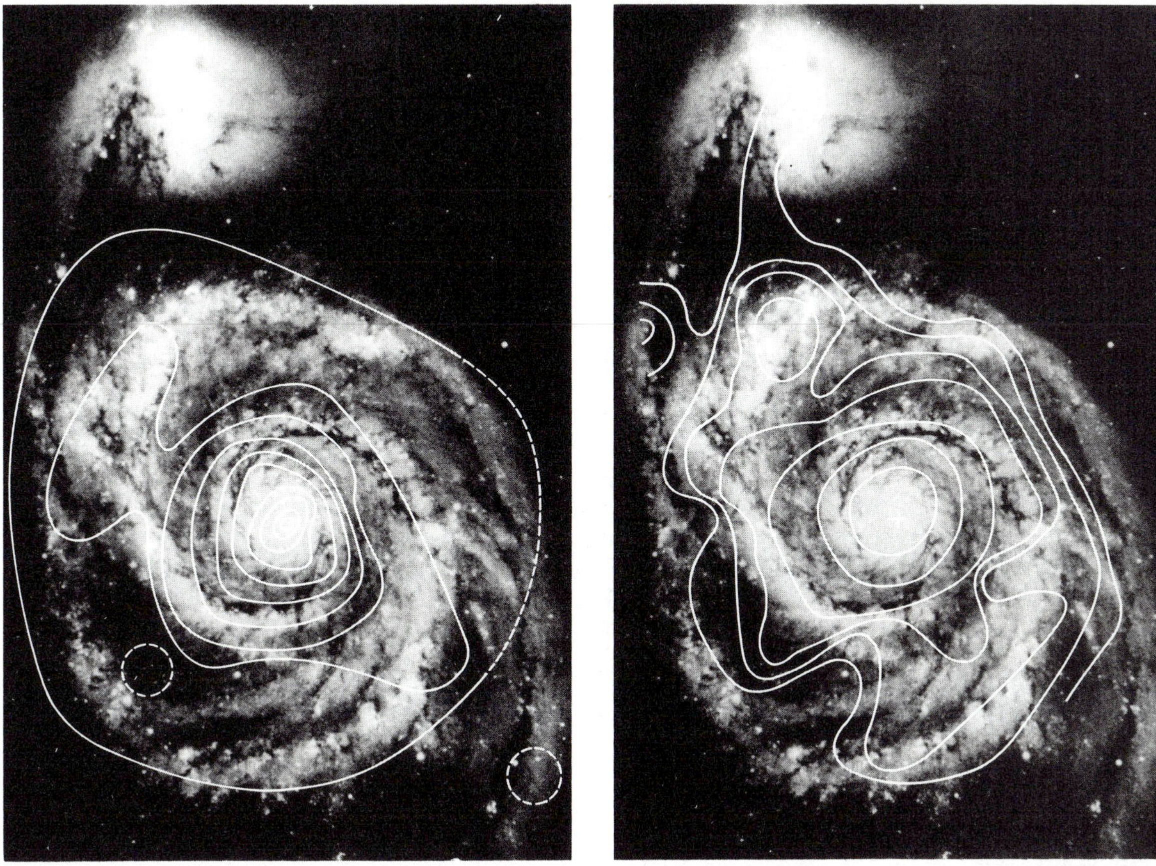

Figure 5.6 DISTRIBUTION OF MOLECULAR CLOUDS in M51 (*left*) is strongly correlated with the distribution of far-infrared radiation from young stars in the same galaxy (*right*). The contours in both cases indicate increasing concentration toward the galactic center. The similarity between the two distributions suggests that the young stars are formed in the clouds and that the rate of star formation is directly proportional to the amount of molecular gas at each point. The molecular-cloud map is based on carbon monoxide data obtained by the authors with the 14-meter radio telescope in Massachusetts. The far-infrared map was made by James Smith and colleagues with the aid of NASA's C-141 airborne observatory. Note the lack of any strong correlation with the galaxy's spiral arms.

yet exist, but an approximate measure of the rate of star formation is obtained from the blue starlight in the galactic disk. This light is generated mainly by hot stars less than a few billion years old. For example, in the late-type spirals IC 342 and NGC 6946 (both classified as *Scd*) we have found similar variations with radius for the carbon monoxide emission and the blue light.

Since the carbon monoxide emission traces the distribution of the densest component of the interstellar medium, it is interesting to compare it with the emission from atomic hydrogen, which presumably traces the less dense (but still neutral) gas. Detailed studies of the atomic-hydrogen content and distribution in external galaxies have been made by many astronomers in the past 10 or 20 years. In spite of the wide range of luminosities in the *Scd* galaxies, all of them have similar distributions of atomic hydrogen, namely a fairly constant density throughout most of the disk except at the center, where there is a deficiency of H I.

In general the size of the atomic-hydrogen envelope for each galaxy is also much larger than the visible galaxy. The carbon monoxide content and distribution for the same galaxies are remarkably different, displaying little resemblance to the atomic-hydrogen profiles, as can be seen in Figure 5.7. Specifically, the galaxies with a high luminosity were found to be abundant in molecular clouds, whereas those with a low luminosity have only a small amount of molecular gas. The molecular distributions also exhibit steep gradients in the radial direction. Thus the high-luminosity galaxies have primarily molecular gas in the center and atomic gas in the outer regions, whereas the low-luminosity galaxies have mostly atomic gas throughout. In this respect our galaxy corresponds more closely to the high-luminosity external galaxies.

The expectation, of course, is that by looking at other galaxies we can come to understand features of our own galaxy. In view of the fact that our galaxy is intermediate along the sequence of spiral types, it was initially most surprising that none of the galaxies first observed for carbon monoxide showed a central peak and ring of molecular clouds like those in our galaxy. It is now apparent that this finding was at least partially attributable to observational selection: most of the galaxies first observed for carbon monoxide were dusty *Sc* spirals, not intermediate types such as our galaxy. As more *Sa* and *Sb* galaxies are observed the ring is seen more often. The origin of the structure might be related to the size of the central, nuclear bulge of old stars.

In fact, if one compares the molecular distributions observed in all the spiral galaxies, their outer disks exhibit a rather similar, smooth decrease in concentration; the dissimilarities appear in the inner disks. In some of the galaxies there is a drop in the density of molecular hydrogen at the center; in others the density of the gas simply continues its increase from the outer disk all the way to the nucleus. This observation suggests that the significant feature is not the peak of the ring but the void that sometimes appears in the inner zone. A possible link between the size of the central-bulge star population and the existence of a breach in the gas distribution might materialize if the gas originally in this zone were exhausted at an early epoch to form the stars in the massive central bulge. Why some galaxies form a more massive bulge than others is a mystery of the initial constitution of galaxies.

The general correlation of the carbon monoxide distribution and the optical luminosity, discovered by us originally in a few selected late-type spiral galaxies, also holds when one compares a sample of galaxies over a wide range of luminosities, all in a given morphological class. The results for the central regions of the type-*Sc* galaxies are quite striking: the optical luminosity varies in direct proportion to the amount of molecular gas. In other words, if more molecular clouds are present, more stars will form and a galaxy will have a correspondingly higher luminosity. This correlation is precisely the same as the one observed in individual *Sc* galaxies such as M51, where the distribution of carbon monoxide emission was found to mimic that of the young stars. The fact that the correlation exists both within particular *Sc* galaxies and between galaxies, however, indicates it is a general feature of star formation in galaxies of this type. If one were to compare the central atomic-hydrogen contents with the optical luminosities within the same sample of galaxies, one would find no correlation at all.

Perhaps the most enigmatic galaxies are the 10 percent or so whose central regions show energetic activity in the form of X-ray and radio emission. At a very low level of activity the center of our own galaxy shows most of these symptoms. For the more extreme cases, such as the quasars, it is believed the activity is generated by an extraordinary object such as a massive black hole at the center of the galaxy. For other cases it has been suggested that the activity is attributable to a burst of star formation. In the latter instance the massive young stars (up to a billion solar masses in the most extreme examples) will have an enormous luminosity for a brief period of about 10 million years; hence a burst of star formation could account for a high instantaneous energy output. Moreover, the supernova explosions coming at the end of the burst would generate high-velocity gas motions in addition to intense X-ray and radio emission. Clearly the duration of the burst is limited by the reservoir of interstellar matter near the center of the galaxy, since most of the gas will eventually coalesce into stars and the subsequent rate of star formation will decrease.

One of the nearest examples of these phenomena is the irregular galaxy M82. For many years it was believed the nucleus of this galaxy had exploded, because plumes of high-velocity ionized gas could be detected above and below the galactic disk (see Figure 5.8). The filaments point radially away from the center of the galaxy just as though they were an outflowing gas stream. The prevailing view now is that many of the peculiarities observed outside the

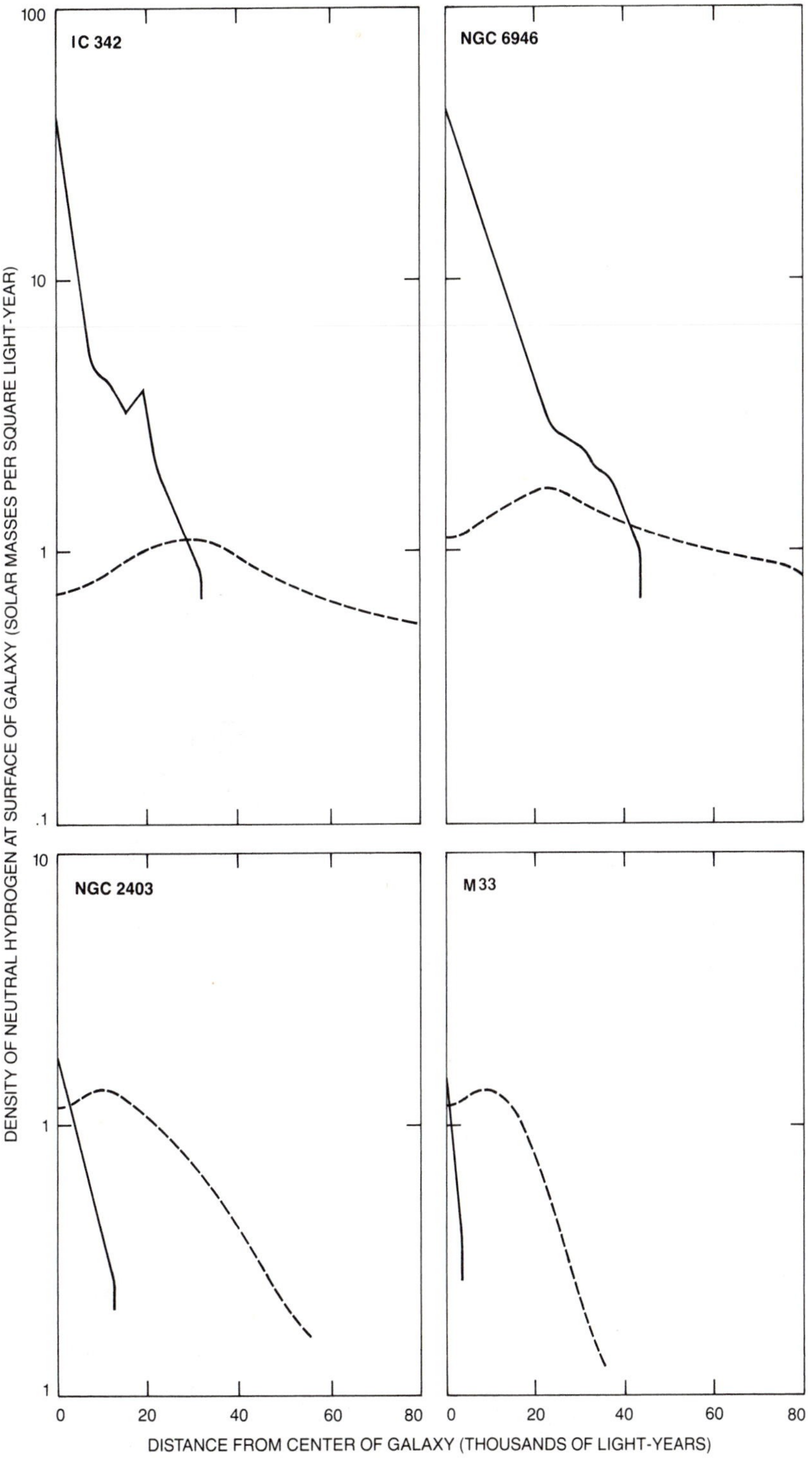

Figure 5.7 COMPOSITION OF INTERSTELLAR MATTER in four *Scd* galaxies is closely related to the luminosity of the galaxy. Near the centers of the two high-luminosity galaxies, IC 342 (*top left*) and NGC 6946 (*top right*), the ratio of molecular hydrogen, or H_2 (*black curve*), to neutral atomic hydrogen, or H I (*broken black curve*), is roughly 100 to one, whereas in the low-luminosity galaxies, NGC 2403 (*bottom left*) and M33 (*bottom right*), the amounts of molecular and atomic hydrogen are about the same. All four galaxies have similar peak values of neutral hydrogen at their surface, so that it is the amount of molecular gas that varies from galaxy to galaxy. (Observations of atomic hydrogen made by David H. Rogstad and G. Seth Shostak.)

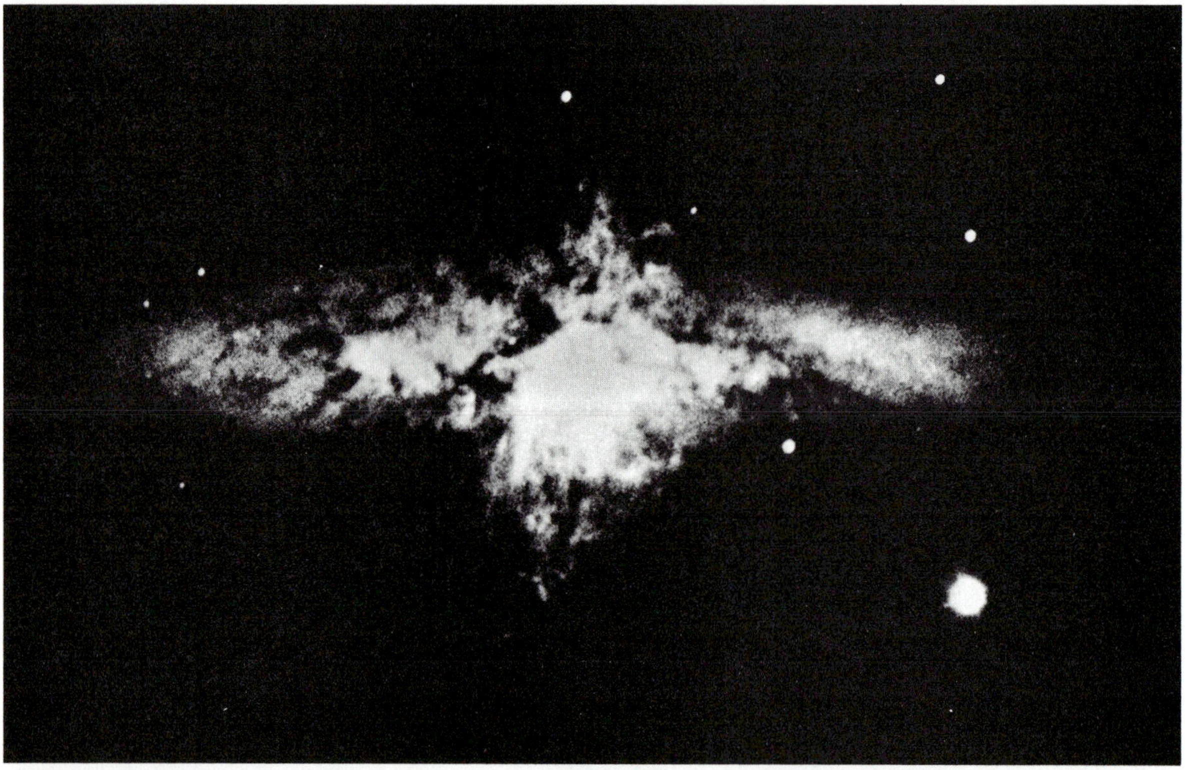

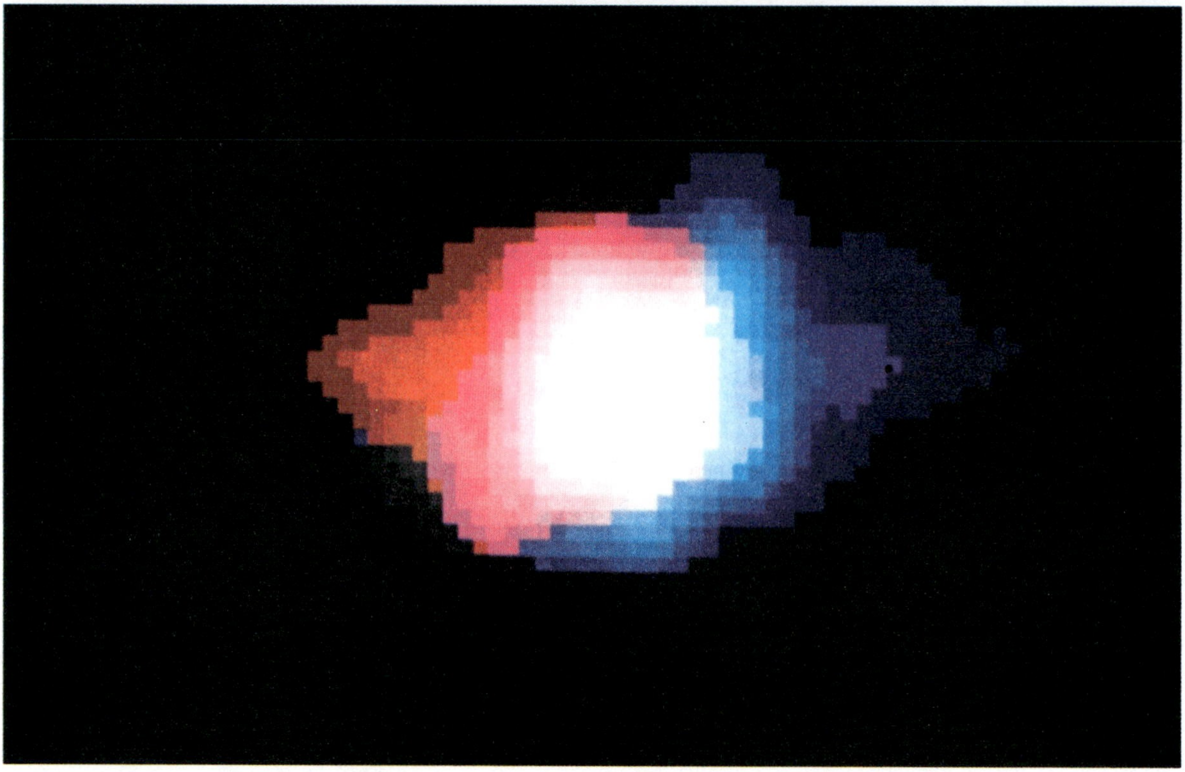

nucleus of M82 can be attributed not to gas flowing out of the galaxy but to intergalactic gas falling into it. The bridge of hydrogen extending across the sky to the nearby spiral galaxy M81 suggests that the infalling gas may have been pulled out of the outer halo of M81 in the course of a close encounter.

M82 is one of the brightest infrared sources beyond our galaxy. At its center the infrared luminosity is about 20 billion times the luminosity of the sun. Because of the great quantities of absorbing dust seen in the plane of the galaxy, the optical luminosity in the same region is a factor of 20 less. Not surprisingly, M82 has been found to be one of the brightest sources of carbon monoxide emission beyond our galaxy. It was in fact the first galaxy detected by Rickard and his colleagues on the basis of its carbon monoxide emission. An analysis of the motions of the molecular gas based on the Doppler effect on the emission reveals a structured pattern, with the receding gas to the north and the approaching gas to the south. Contrary to the situation in normal spiral galaxies, where the axis of rotation is perpendicular to the galactic disk, here one finds an axis tilted at nearly 45 degrees to the disk. This finding suggests the presence of large motions directed radially toward or away from the center of M82; in other words, the gas clouds have velocities significantly different from those of normal circular orbits. These peculiar motions are perhaps the consequence of an infall of gas from outside the galaxy.

Further study of the carbon monoxide emission from M82 shows that the greatest concentration of molecules is at the center of the galaxy near the peak in the far-infrared emission. The density of the molecular gas exceeds that of atomic hydrogen over the entire optical disk and well up into the vertical filaments. Indeed, the total mass of molecular gas in the galaxy is almost half the total mass of the stars, or roughly three times the fraction observed in normal spiral galaxies. In many respects the peculiarities observed in M82 bear a strong resemblance to what one would expect of a young galaxy: a stellar population with a high proportion of massive, short-lived stars, an abundance of interstellar gas not yet coalesced into stars and continuing infall of intergalactic gas from the outer part of the protogalactic cloud. Could we be witnessing here the birth —or, more likely, the rejuvenation—of an entire galaxy?

Figure 5.8 M82 OBSERVED at two different wavelengths. The photo at top was made in red light with the 200-inch telescope on Palomar Mountain. It shows a generally amorphous disk with dark dust lanes silhouetted against its surface and a tangled array of filaments extending outward at right angles from the center of the disk. The brightness of the image at bottom, based on the 2.6-millimeter emission from carbon monoxide molecules in the same region and made by the authors, is proportional to the intensity of the emission. The colors represent the velocity of the gas with respect to the center of the galaxy; red is gas that is moving away from the viewer, blue, gas that is approaching.

In some galaxies, particularly those with a fairly modest rate of star formation in the central region, the activity in that region might be initiated and sustained by the infall of gas from the outer disk of the galaxy. A good example of the phenomenon is presented by the galaxy IC 342. Kwok-Yung Lo and his colleagues at Cal Tech have recently mapped the carbon monoxide emission in the nucleus of this galaxy with the new high-resolution, millimeter-wave interferometer at the Owens Valley Radio Observatory. Their data clearly show for the first time that the molecular clouds in IC 342 are confined to a barlike structure about 6,000 light-years long (see Figure 5.9). The measured velocities of the clouds suggest a significant infall toward the center of the galaxy where previous infrared observations had indicated a higher-than-normal rate of star formation. The bar may act to channel gas clouds into the central zone.

At an even higher level of activity is the spiral galaxy NGC 1068. It is in a class of galaxies distinguished by their extraordinarily bright and compact optical nuclei, which exhibit strong emission lines from high-velocity ionized gas. The far-infrared luminosity is a staggering 200 billion solar luminosities, all originating from the central 5,000 light-years. This galaxy is also extremely rich in molecular gas, as can be judged from the strong carbon monoxide emission. Nevertheless, it is clear that the current level of activity can be sustained for no longer than a few hundred million years if the luminosity is produced by young stars and the supply of star-forming material is limited to what is now seen in the central regions. It is difficult to understand how the supply in the central region can be replenished by transport from the outer disk, where the gas is currently in a circular orbit. Thus bursts of star formation have usually been favored to explain these galaxies, and activity in the central region is probably sporadic.

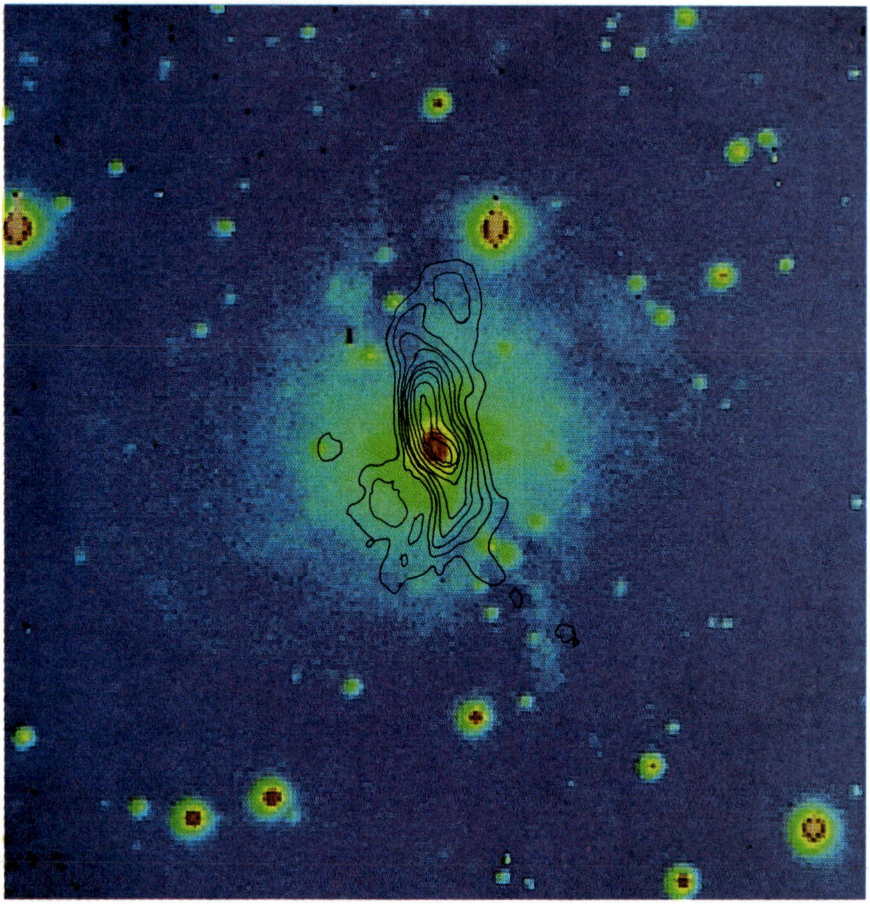

Figure 5.9 MOLECULAR CLOUDS in the nucleus of IC 342 are confined to a bar-shaped region measuring about 6,000 light-years in length (*black contours*). The data suggest that the clouds are flowing in toward the center of the galaxy along the bar, possibly accounting for the high rate of star formation there. The molecular clouds were plotted on the basis of their 2.6-millimeter radio-wave emission from carbon monoxide molecules. The false-color optical image was obtained with a charge-coupled device attached to the 60-inch telescope on Mount Wilson.

One of the most intriguing questions for the starburst models is what regulates the process. How does the burst start, why does it spread and how is the fire quenched? It may be that the activity in the molecular clouds distributed throughout the inner galactic disk is triggered by activity in the central object: a black hole, if one exists. NGC 1068 appears to show both kinds of activity. A source less than 100 light-years in diameter has an infrared, optical and X-ray luminosity of nearly 100 billion suns. Presumably this energy originates in a central compact object. At the same time there is clear evidence that the far infrared luminosity of a similar magnitude originates in a disk of clouds extending over at least a few thousand light-years. Perhaps the central source stimulates star formation in the outlying clouds by sending explosive shock waves through the galaxy. Alternatively it could be that the region is inert until a critical mass of star-forming material accumulates; then a small initial spurt of star formation might not only become self-sustaining but also lead to a runaway chain reaction of massive-star formation. At present the molecular observations cover too few galaxies for one to tell if there are some galaxies with a large abundance of gas but little active star formation, as would be suggested by the second possibility. In addition the second possibility leaves unanswered the question of the nature of the central object.

It is now well established that most stars must form in molecular clouds. This conclusion follows from studies of nearby star-forming regions such as the Orion nebula; it also is implied by the strong correlation found in the external galaxies between the molecular gas and the radiation from young stars. The other major gaseous component of these galaxies, atomic hydrogen, does not show this correlation.

In spite of the understanding gained in the past few years, astronomers are still left with perplexing questions about the nature of the spiral arms in galaxies. What is this phenomenon that dominates the visible morphology of galaxies? Is an arm merely a phase change in the galactic disk, like the puffy white clouds in the earth's atmosphere, or is an arm more substantial in structure, perhaps a wave in the density of matter that propagates through the galactic disk? The answer has been elusive partly because the nature of the arms in one galaxy, or even in one region of a galaxy, may differ from that in another.

The observational studies of molecular gas have clearly demonstrated that the abundance of molecular clouds can vary greatly among galaxies and at different radii within the same galaxy. If the existence of molecular clouds is a prerequisite for star formation, then in galaxies with large amounts of molecular gas that condition will be satisfied throughout the disk. In such cases one expects the patterns of star formation to be much less ordered and coherent, since star formation will be widespread. On the other hand, for those galaxies where the amount of molecular gas is low, the condition may be satisfied only in certain places. In these latter galaxies it then becomes much easier for a grand spiral pattern of star formation to arise with molecular clouds only along the arms. Clearly in analyzing the large-scale patterns of star formation one must consider separately those galaxies with an abundance of molecular gas, such as our galaxy, and those with a small amount, such as M31 (the Andromeda galaxy) or M81. Those two galaxies are often cited as examples of the kind of grand spiral pattern consistent with the density-wave theory; they are also both extremely deficient in molecular gas. The low abundance of molecular gas is consistent with the finding by Richard A. Linke and Anthony Stark of Bell Laboratories that the carbon monoxide emission in M31 is closely confined to the visible spiral arms.

The major problem in understanding spiral structure is not how a curved arm is formed but rather how the structure is maintained over a long time and becomes symmetrical about the center of the galaxy. It is well known that any disturbance moving radially in the disk of a spiral galaxy will be forced into a spiral simply as a result of the galaxy's rotation. Typically it is found that the speed of rotation in a galaxy is fairly constant with radius over most of the disk, implying that matter orbiting at a large radius will lag and matter orbiting at a small radius will advance (with respect to a rotating point of origin midway between them). Thus in these galaxies any effect such as an explosion or a rash of star formation spreading from one cloud to another would naturally form a curved arm. The difficulty arises when one tries to coordinate the phenomenon over the entire disk.

It now seems possible that much of the confusion on this subject may be the result of attempting to force the coherence and symmetry of a spiral pattern on all galaxies rather than simply accepting the fact that there are some galaxies, perhaps a majority of them, where there is little evidence of coherence. In this picture the galaxies with a truly coherent spiral pattern would be those such as M31 and M81, which have a comparatively low abundance of star-forming clouds, or those such as M51, which has a companion galaxy sufficiently close to exert a major tidal pull over the entire galactic disk. In the remaining galaxies with an abundance of clouds the formation of stars will be more widespread and less organized.

One indisputable fact is that the arms, such as they are, are places where massive stars form. Whether there is an enhancement in the rate of formation of all stars in the arms or just a shift in the relative proportion of massive stars is not known. Within the next decade one can look forward to the answer to this important question, since satellite-borne infrared telescopes aimed at the external galaxies will have enough angular resolution to clearly separate their arms and the regions between the arms.

Through observations of the large-scale distribution of molecular clouds, both in our galaxy and in other galaxies, understanding of the relations between these cold, dense regions and the global properties of galaxies such as morphology and luminosity has been deepened. Since the molecular clouds are the precursors of star formation, it is now possible to obtain a wealth of information on a crucial phase in the life cycle of stars and to begin to understand the evolution of galaxies.

It has been supposed that most of the organic matter in the universe (that is, carbon compounds more complex than carbon monoxide) must be on the surface and in the atmosphere of planetary bodies. The enormous mass of the molecular clouds implies, however, that they are the principal reservoirs of organic matter. Moreover, if the solar system were to pass through one of these clouds, the absorption of light would be so great that all but the few nearest stars would disappear from view. If the sun traveled with a typical stellar velocity of 20,000 miles per hour, more than two million years would elapse before the earth emerged from the gloom. Given the abundance of the clouds in our galaxy, such an event must come roughly once every billion years, or about five times in the history of the earth. If human beings had evolved during such an episode, both their vision of the universe and their philosophical outlook would have been fundamentally different.

The Coronas of Galaxies

6

Satellite observations indicate that the entire disk of our galaxy has an extended envelope of hot gas. The same appears to be true of other spiral galaxies.

• • •

Klaas S. de Boer and Blair D. Savage
April, 1982

The familiar picture of a spiral galaxy is a thin disk of stars, gas and dust with a somewhat thicker central bulge. Our own galaxy is such a spiral system, with a disk some 100,000 light-years in diameter. Surrounding the disk is a region called the halo, thinly populated by old stars and globular star clusters. The disk of the galaxy has been patiently mapped by means of radiation emitted by the gas in it, and the halo has been mapped by means of radiation emitted by the stars in it. There is a third component of the galaxy that is difficult to detect from the earth. Called the galactic corona, it is an envelope of hot gas that extends tens of thousands of light-years on each side of the central plane of the galactic disk.

Given the difficulty of detecting the corona from the earth, how was it discovered and how is it studied? The answer is that the hot gas of the corona absorbs ultraviolet radiation. Therefore if the gas lies between an ultraviolet-emitting star and an ultraviolet-detecting instrument, the gas can be detected by its absorption of the star's ultraviolet radiation. Most ultraviolet wavelengths, however, cannot penetrate the earth's atmosphere, so that such observations can be made only from above the atmosphere. This was one of the reasons the satellite *International Ultraviolet Explorer* (*IUE*) was launched in 1978. Since that time the instruments of the satellite have begun to yield a rich picture of the galactic corona and its dynamic physical processes (see Figure 6.1).

Until about 20 years ago it was believed that much of the volume on each side of the central plane of the galaxy held no gas. It is now thought that in this volume are gigantic streams of gas ascending from the galactic plane and descending toward it. The force that drives the streams is the explosion of supernovas in the galactic plane. Such an explosion heats the surrounding gas to a temperature of a million degrees Kelvin (degrees Celsius above absolute zero). The hot gas moves outward from the galactic plane, cooling as it does so. It then begins to condense and move back toward the plane of the galaxy, where the process may begin again. The entire cycle has been called the galactic fountain.

In addition to securing the first direct evidence for the corona of our galaxy the instruments of the *International Ultraviolet Explorer* have probably detected coronas around two neighboring galaxies. Indeed, coronas of hot gas may be common in the universe. If they are, they may provide a simple explanation for a long-standing problem about the light of the enigmatic quasars.

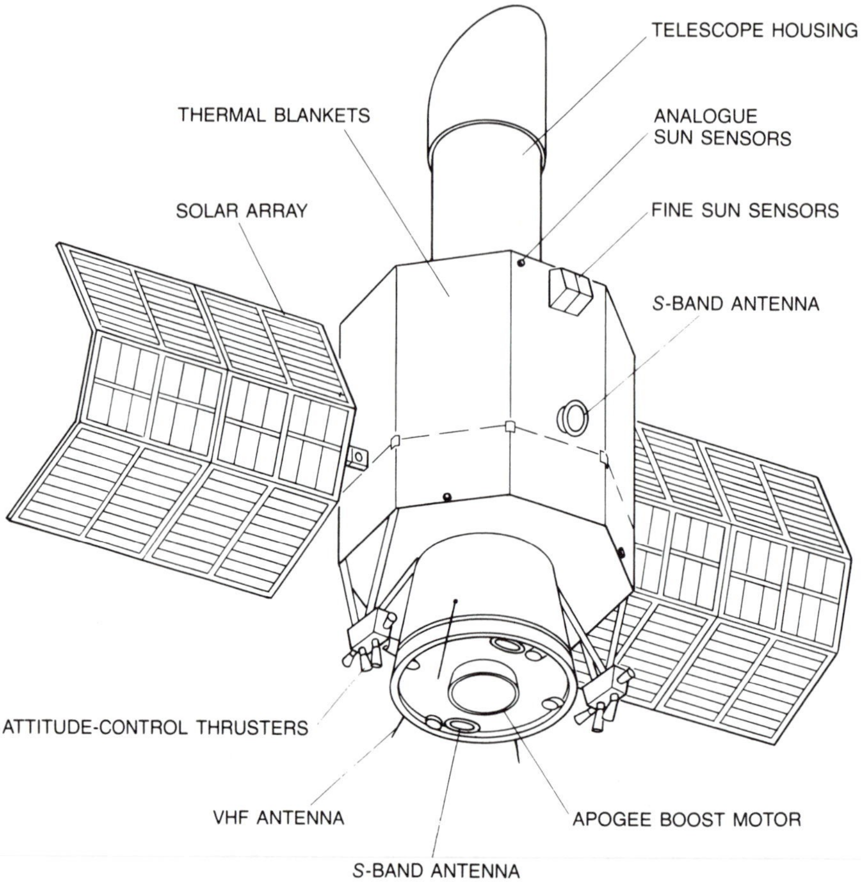

Figure 6.1 INTERNATIONAL ULTRAVIOLET EXPLORER (IUE) carries a telescope and spectrograph. With the telescope pointed at a distant star the spectrograph records the intensity of ultraviolet radiation, including specific wavelengths absorbed as the radiation from the star passes through the intervening gas. Such data can indicate the gas's composition, temperature and velocity. The solar panels provide the power for the satellite's instruments. The sun sensors make it possible to adjust the satellite's attitude with respect to the sun. VHF antenna receives commands from ground stations; *S*-band antenna relays data to the ground.

The galactic corona has been named by analogy with certain other phenomena. "Corona" is from the Latin for "crown" or "garland." In astronomy the word has been employed to describe the extended envelope of gas that surrounds some celestial objects and also to describe the light emitted or scattered by such gas. The most familiar corona is of course the corona of the sun. The solar corona cannot be seen with the unaided eye except during a total eclipse of the sun, when it appears as a luminous halo around the dark eclipsing disk of the moon. The light that appears to originate in the solar corona is mainly the light of the sun's luminous disk, scattered by free electrons in the corona's diffuse gas.

The spectrum of the radiation emitted by a celestial object is a signature of the object's temperature. The spectrum of the solar corona shows that the corona is extraordinarily hot: between one and two million degrees K. (The temperature of the sun's visible surface is 6,000 degrees.) Observations made with satellite instruments capable of detecting X rays indicate that many stars have hot coronas much like the corona of the sun. It has also been shown that planets and comets have coronas. The coronas of planets and comets, however, are much cooler than those of stars, having a temperature of at most a few thousand degrees K.

The evidence that has been gathered for the existence of a corona around the galaxy has created a problem of terminology for astronomers. The trouble arises from the fact that the meaning of the word corona overlaps that of the word halo. "Halo" is from the Greek *halos,* which originally meant the disk of the sun or the moon. Aristotle used the word in this sense; it was only later that it came to mean the nimbus around the head of a saint. In meteorology "halo" refers to the ring of scattered light around the sun or the moon when they are seen through cirrus clouds.

In astronomy the meaning of the word halo has gradually become limited to the region around the galactic disk populated by old stars and globular clusters. When it was concluded that this spherical region might contain gas, the envelope of gas was called the gaseous halo by some astronomers and the corona by others. Here we shall reserve "halo" for the population of old stars and "corona" for the gaseous envelope of the galaxy.

The disk of a galaxy such as ours has spiral arms that appear much brighter than the regions between them. The concentration of stars in the spiral arms is not, however, as high as it might seem; the reason is that the arms contain the hottest and most luminous stars. Such stars are short-lived, with a life span of from one to 10 million years. They make up only a small minority of the stars in the galaxy. Most stars, the sun among them, are less brilliant, have lifetimes of billions of years and are uniformly distributed in the galaxy.

Much of the gas in a spiral galaxy is concentrated near the central plane of the galactic disk. The gravitational forces exerted by the stars in the central plane tend to hold the gas in it. The thickness of the layer of gas is less than 1 percent of the diameter of the disk.

Although the disk of interstellar gas occupies a small volume compared with the volume of the galaxy as a whole, it is the scene of much activity of astronomical significance. One current hypothesis is that the gas of the disk is itself compressed into spiral arms by the gravitational forces exerted by all the rotating material in the disk. If the density of the gas in a particular region of the disk gets high enough, the mutual gravitational attraction of its own atoms or molecules may cause it to begin to coalesce. The process can result in the formation of the dense dark clouds in which stars originate. The gas can also be compressed by the blast wave from the explosion of a supernova: a massive star at the end of its evolutionary development. Furthermore, the gas of the disk has a magnetic field that may serve to maintain the spatial relations among some of the structures of the galaxy. The magnetic field may have loops that extend away from the central plane and into the galactic corona.

It was not until the early 1950's that the extent and significance of the gas in the central disk began to be appreciated. The best evidence for its existence came from observations made with radio telescopes. Most of the disk gas is hydrogen, roughly half in the atomic form (H) and half in the molecular (H_2).

Like the electrons of other atoms, the single electron of a hydrogen atom is not free to move randomly around the atomic nucleus; it must remain in certain strictly specified patterns or states. The states differ according to the amount of energy the electron possesses in each one. The transition of an electron from one state to a state of higher energy requires that a specific quantity of energy be put into the atom; if the electron reverts to the original state, the same quantity of energy is released.

The energy needed to change the state of an electron may come from one of several sources, including collisions between atoms and electromagnetic radiation from nearby stars. If the energy is in the form of radiation, a particular wavelength is required. The reason is that the energy needed to change the state of the electron is rigidly specified. The energy of any form of radiation is inversely proportional to its wavelength; the higher the energy, the shorter the wavelength. Therefore only radiation of a certain wavelength can accomplish a particular electron transition. Since the same quantity of energy is released when an electron returns to its initial state, the emitted radiation will have the same wavelength as the radiation that caused the transition. Because a particular atom is capable of only certain transitions the wavelengths that are absorbed or emitted as radiation passes through a cloud of gas yield information about the atoms making up the gas.

In the case of the cool gas in the galactic disk the energy needed to change the state of the electron is furnished by collisions between hydrogen atoms. Because of the low temperature of the gas such

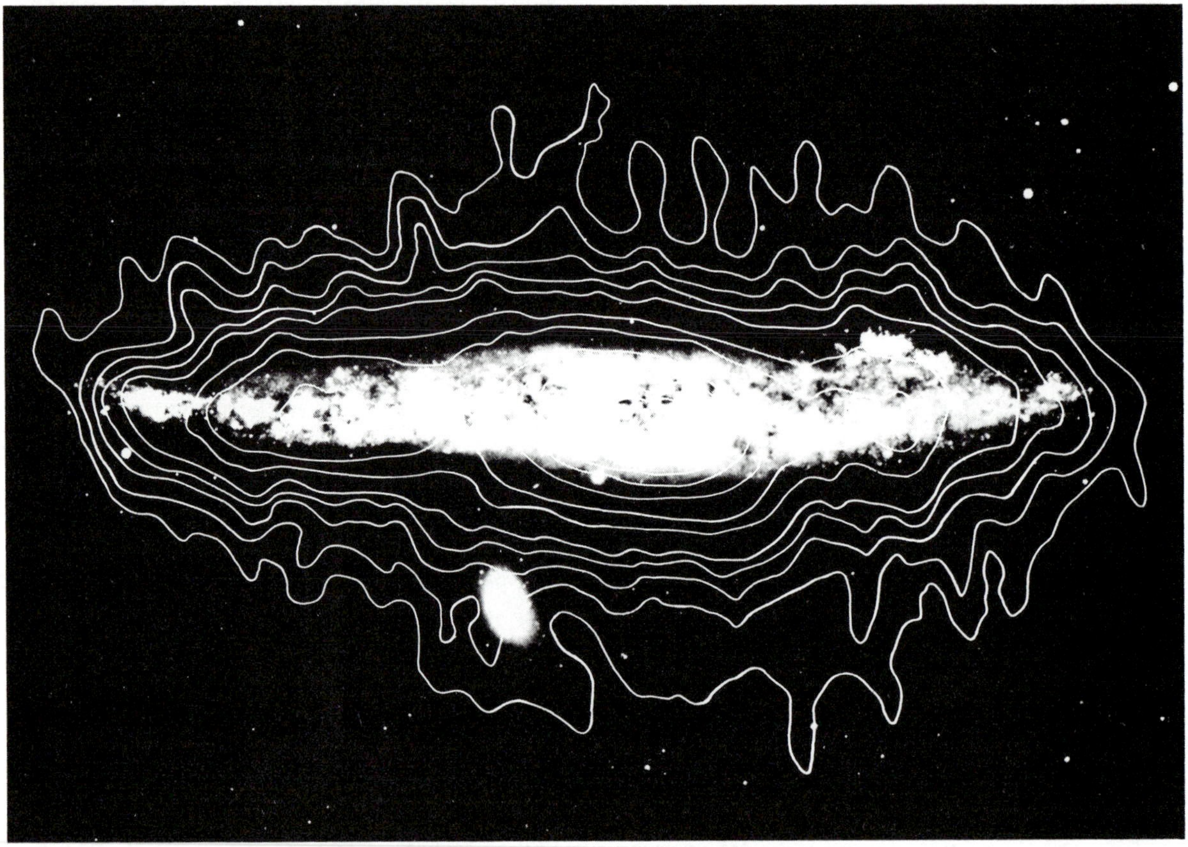

Figure 6.2 RADIO-EMISSION MAP superposed to scale on a photograph of spiral galaxy NGC 4631 seen edge on is indirect evidence for the existence of a corona of hot gas around the galaxy. The radio emission, mapped at a frequency of 1,412 megahertz by Renzo Sancisi of the Kapteyn Laboratory and his colleagues with the Westerbork Synthesis Radio Telescope, is not from gas atoms but from free electrons spiraling around lines of force in the galaxy's magnetic field. The radio intensity decreases with distance from the central plane of the galaxy; the contours extend 30,000 light-years from the plane.

collisions are followed by only the lowest-energy electron transition. Therefore when the electron reverts to the lower state the radiation emitted has a long wavelength: 21 centimeters, which is in the radio range.

With radio telescopes 21-centimeter radiation has been exploited to detect disk gas at large distances from the sun, which is about 25,000 light-years from the center of the galaxy. Hydrogen has been detected at distances as great as 100,000 light-years. Since the disk is about 100,000 light-years across, such a distance implies that the gas is present throughout the disk.

Furthermore, the radio-telescope observations indicate that the gas of the disk participates in the rotation of the galaxy about an axis perpendicular to the central plane of the disk. The rotation is such that matter more than 10,000 light-years from the center of the galaxy moves in orbit with a velocity of about 220 kilometers per second. Given the diameter of the galactic disk, gas at the periphery of the disk may have revolved about the galactic center only some 25 times since the origin of the galaxy some 10 billion years ago.

Similar principles have been employed in the investigation of components of the disk gas other than hydrogen. When a telescope is aimed at a distant star, the radiation emitted by the star must pass through the gas in order to reach the telescope. The emission from the star generally consists of radiation of many different wavelengths, and as the radiation passes through the gas specific wavelengths will be absorbed. As we have seen, the wavelengths that are absorbed are those with precisely the quantity of energy needed to raise the electrons of the atoms of the intervening gas from one state to another. In the spectrograph the absorbed wavelengths are recorded as dark lines, representing sharp decreases in the radiation in the spectrum of wavelengths.

Absorption lines can yield information not only about the composition of the interstellar gas but also, by taking into account the Doppler effect, about its velocity. If the gas is moving away from the observer, the absorption occurs at a slightly longer wavelength. This "red shift" is directly proportional to the velocity of the gas along the line of sight, and so the velocity can be readily calculated.

The pattern of the absorption lines indicates that the interstellar gas includes many elements besides hydrogen. Some elements are found primarily in the solid particles of interstellar dust. The relative proportions of the elements in the gas and the dust combined are probably similar to those in the atmosphere of the sun. (The sun's atmosphere is mostly hydrogen, but it includes more than 50 other elements.)

Each atom in the interstellar gas can assume several different states of ionization. An ion is an atom that has more or fewer electrons than the atom in the neutral (un-ionized) state. Most of the ions in the interstellar gas have had electrons removed and are therefore positively charged. Electrons can be removed from an atom by the same processes that raise them from one energy state to another, namely collisions with other atoms and exposure to electromagnetic radiation.

The degree of ionization depends on how many electrons have been removed. For example, if a single electron is removed from a neutral interstellar carbon atom, the atom acquires one unit of positive charge and is said to be once ionized. If another electron is removed, the atom, with two units of positive charge, is twice ionized. In atomic spectroscopy the neutral carbon atom is designated C I, the once-ionized C II and the twice-ionized atom C III. Much of the hydrogen in the interstellar gas is H I; the large clouds of neutral hydrogen between stars are known as H I regions. Near a hot star, however, energetic photons (quanta of radiation) remove the electron from the hydrogen atom; the result is the formation of an H II region.

The elements in the interstellar gas that are heavier than hydrogen are also influenced by radiation from nearby stars. For each of these atoms a particular quantity of energy is required to remove an electron; the quantity is referred to as the ionization energy. The differences in ionization energy result in varying degrees of ionization of the heavier elements under the conditions in the interstellar gas.

As in the neutral atom, each electron in an ion moves only in strictly specified ways. The wavelength of the radiation absorbed in the electron transitions can yield considerable information about the ions. Work done in the 1950's by Guido Münch of the California Institute of Technology with the 200-inch Hale telescope on Palomar Mountain yielded significant results on the interstellar gas. Münch directed the telescope at stars far above the galactic plane. Examination of the resulting spectra indicated absorption by Na (sodium) I and Ca (calcium) II. The absorption lines were quite complex; they indicated the presence of gas clouds with high velocities.

The high-velocity clouds observed by Münch have since been studied intensively by astronomers working with the 21-centimeter emission line of hydrogen. The existence of gas clouds at very large distances from the galactic plane raises several problems. The first problem is related to the gravitational forces exerted by the stars of the galactic plane. If a cloud of gas is to rise above the plane, it must overcome these forces. The fact that the high-velocity clouds had risen so far above the galactic plane indicated that they must have had a very high velocity when they left the central disk of the galaxy.

Another problem was somewhat more perplexing. If the space around a high-velocity cloud were empty, the cloud would expand, disperse and disappear. The continued existence of the clouds implies that some force holds them together. In interstellar space the likeliest candidate for such a force is the pressure exerted by gas surrounding the high-velocity clouds. In 1956 Lyman Spitzer, Jr., of Princeton University proposed that the galaxy is surrounded by a corona of hot gas, and that it is the coronal gas that holds the high-velocity clouds together.

Spitzer had inferred the existence of the hot coronal gas from evidence gathered by Münch and others on the high-velocity clouds. In discussing the possibility of proving the existence of the coronal gas in a more direct way Spitzer noted that the presence of the coronal gas would probably be indicated only by the absorption of radiation with a wavelength of 2,000 angstrom units or less, which is in the ultraviolet range.

At the temperature predicted for the gas in the hottest part of the corona most atoms would have lost many electrons; the high temperature of the gas increases the kinetic energy of the colliding atoms and therefore the probability that one of the atoms will lose an electron. The highly ionized atoms include C IV, Si (silicon) IV, N (nitrogen) V and O (oxygen) VI. These ions all show absorption lines at wavelengths shorter than 2,000 angstroms. The absorption lines of the abundant atoms carbon, nitrogen, oxygen, magnesium, aluminum, sulfur, silicon, iron, nickel and zinc in their neutral or once-ionized states are also in the ultraviolet.

At the time Spitzer published his classic paper it was impossible to verify the existence of the corona; the necessary observations would have had to be made from the ground—from the bottom of the earth's ultraviolet-absorbing atmosphere. In order to gather more direct experimental evidence for the existence of the galactic corona an observational platform outside the atmosphere was needed. Earth satellites have supplied such a platform. The first satellites with instruments capable of detecting absorption in the ultraviolet part of the spectrum were launched by the National Aeronautics and Space Administration (NASA) in the early 1970's; they were known as the Orbiting Astronomical Observatories (OAO). The second successfully launched observatory in the series, *Copernicus,* was in operation from 1972 until the end of the OAO program in 1980. *Copernicus* carried experimental apparatus designed specifically for the study of lines in the ultraviolet absorbed by interstellar gas. The *Copernicus* project was directed by Spitzer, who also supervised the design of its instruments.

Of the many findings made with *Copernicus* one of the most important was the observation of O VI absorption in the interstellar medium near stars in the vicinity of the sun; the work was done by Edward B. Jenkins and his colleagues at Princeton University. The presence of O VI indicates that this interstellar gas has an extremely high temperature. William L. Kraushaar and his co-workers at the University of Wisconsin at Madison also concluded that components of the interstellar medium near the sun were extremely hot; their results were obtained by studying the emission of X rays.

The hot components of the interstellar medium, which were discovered in the 1970's, have been the subject of much recent experimental and theoretical work. The best current hypothesis is that the gas is heated by shock waves from the explosion of supernovas. In a spiral galaxy these dramatic explosions come about once every 100 years.

The later work tended to support Spitzer's hypothesis that the high-velocity clouds are held together by pressure exerted by surrounding clouds of hot gas. *Copernicus,* however, did not have instruments sufficiently sensitive to measure the absorption of radiation from distant stars far above the plane of the galaxy. Measurements of this kind are needed to detect the galactic corona. It was not until the launching of the *International Ultraviolet Explorer* in January, 1978, that the corona could be actively studied.

The *IUE* is operated jointly by NASA, the European Space Agency and the British Science Research Council. It has served for observations of many phenomena other than the galactic corona. It is a geosynchronous satellite: its period of rotation about the earth is synchronized with the period of the earth's rotation so that the satellite remains stationary above South America, about 40,000 kilometers from control stations in Europe and the U.S. The satellite carries a telescope and a spectrometer (see Figures 6.1 and 6.3). The beam of ultraviolet radiation from the spectrometer is directed toward television cameras sensitive to ultraviolet radiation; the cameras create an image of the spectrum that is subsequently relayed to the ground (see Figure 6.4).

The resolving power of the spectrometer can be altered by the observer on the ground. In one mode details of the spectrum seven angstroms wide can be detected; in the other mode details .1 angstrom wide can be detected. The high-resolution spectrum is essential for the study of absorption by the galactic corona.

To examine regions of the corona far from the plane of the galaxy we directed the telescope of the *IUE* toward a star in the Large Cloud of Magellan. The Large Cloud and its close neighbor, the Small Cloud of Magellan, are small galaxies near our own: about 160,000 light-years from the sun (see Figure

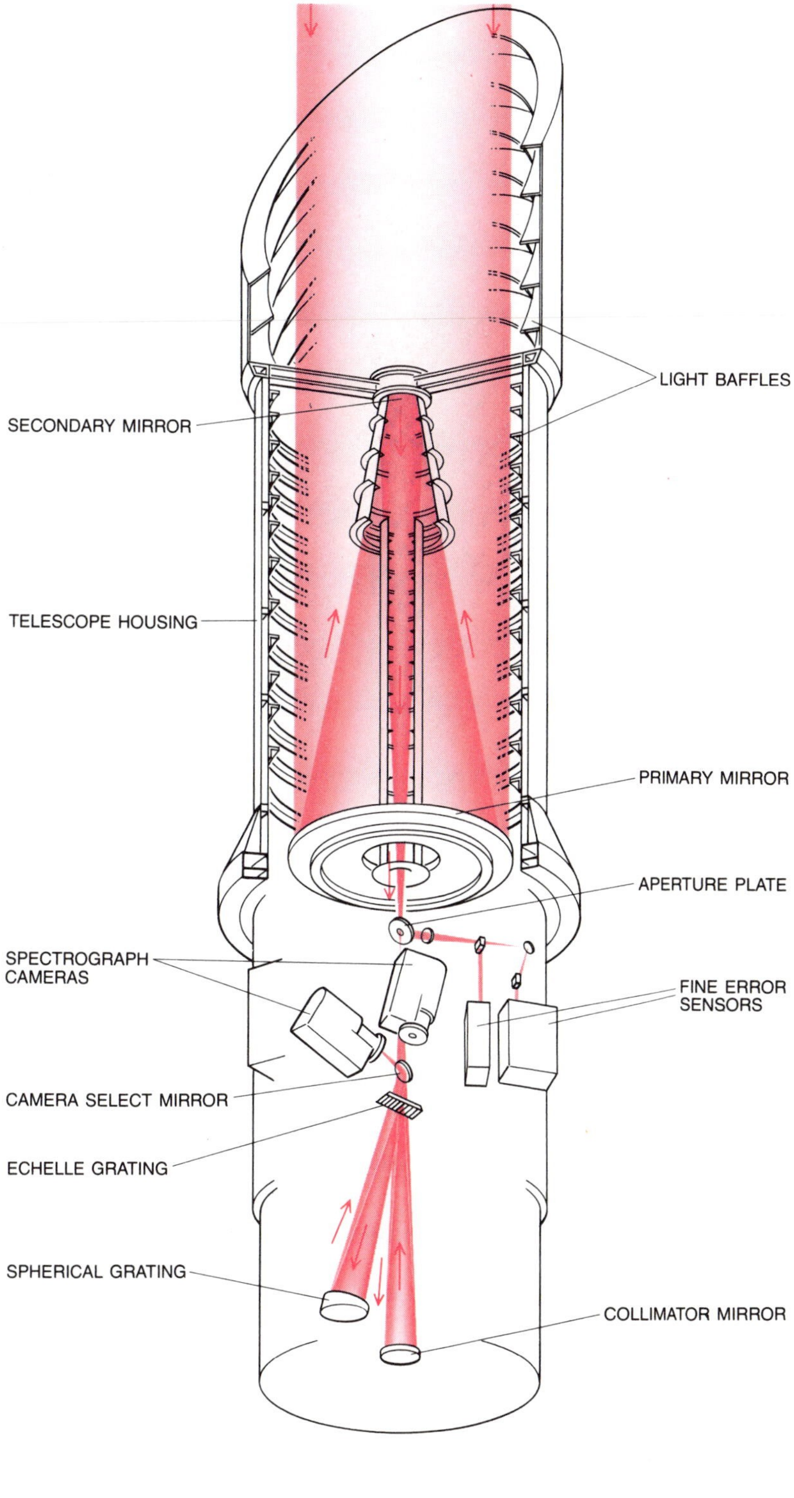

Figure 6.3 INSTRUMENTATION OF THE *IUE*. The primary mirror of the telescope is 45 centimeters in diameter. The radiation that strikes it is reflected to a secondary mirror and back through an aperture in the primary mirror into the spectrograph compartment. Mounted behind the telescope are two spectrographic systems. Most of the beam passes through the aperture plate to the collimator mirror and is reflected onto the echelle grating, which breaks it into its constituent wavelengths. The spherical grating separates the beam into spectral orders and directs it into television cameras that are sensitive to ultraviolet radiation.

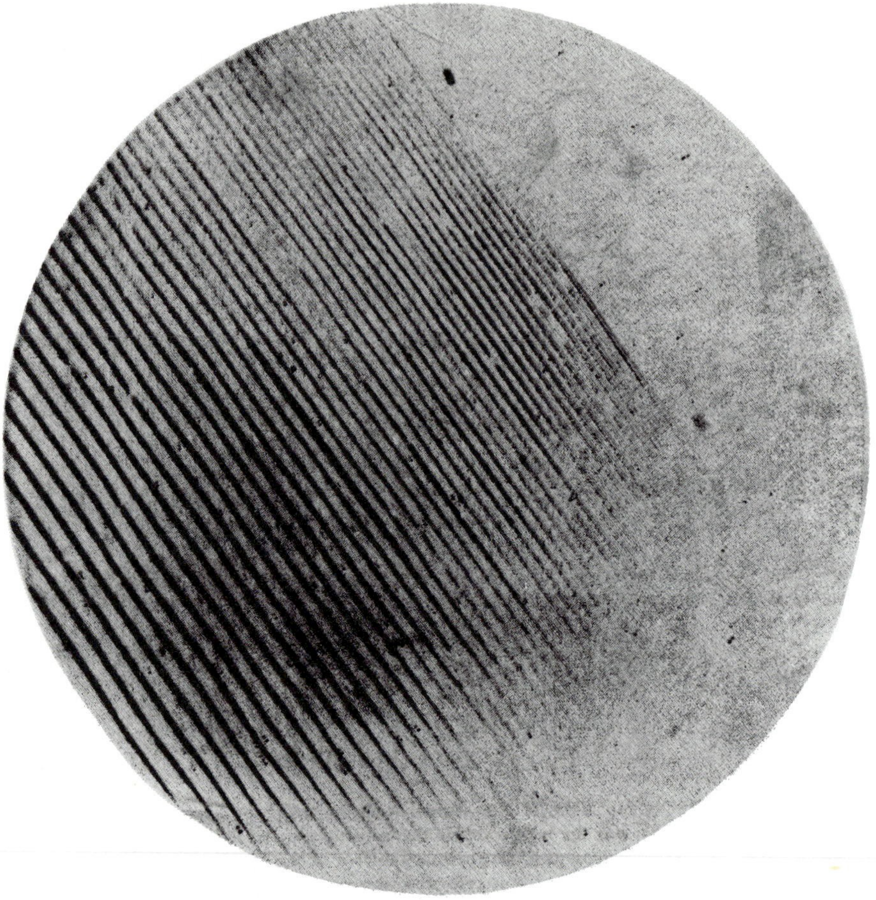

Figure 6.4 ECHELLE SPECTROGRAM made by the *IUE* for the star HD 38282 in the Large Cloud of Magellan. It includes radiation with wavelengths from 1,200 angstroms (*upper right*) to 2,000 angstroms (*lower left*). The gratings of the spectrograph break the radiation down into orders about 15 angstroms wide. Each dark diagonal line in this positive print corresponds to one order. The white space in certain orders indicates that radiation of a particular wavelength has been absorbed by the gas of the corona. The dark spots were made by cosmic-ray particles. The white dots are a grid for correcting the geometrical distortions in the television image.

6.5). At that distance the stars of the Clouds of Magellan are so faint that an exposure of six hours was required to make a high-resolution spectrogram.

The Large Cloud of Magellan has an apparent velocity of 270 kilometers per second away from the sun. Much of the apparent velocity is actually due to the motion of the sun around the center of our own galaxy. When we examined absorption spectra made with the *IUE,* we found a complex pattern giving evidence of absorption by gas clouds with quite different velocities (see Figure 6.6). We found evidence of absorption at a velocity of zero with respect to the sun and also at a velocity of 270 kilometers per second. The absorption lines were clearly due to gas near the sun and to gas in the Large Cloud of Magellan.

Surprisingly, we also observed pronounced absorption by gas with an intermediate velocity. The wavelengths of the absorption implied that the ions responsible were C II, Mg (magnesium) II and Si II; such ions are formed at moderate temperatures. Even more intriguing, however, was the finding of strong absorption by C IV and Si IV. Such ions are

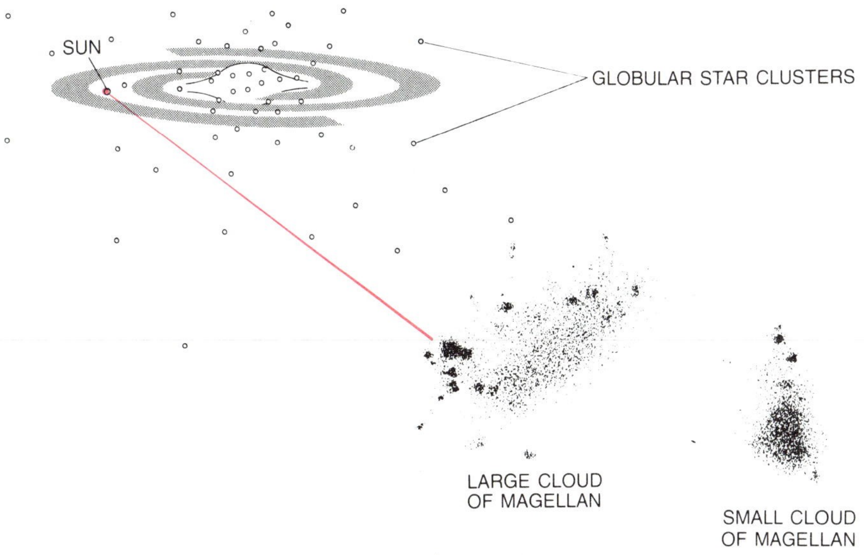

Figure 6.5 CLOUDS OF MAGELLAN are two small galaxies near our own. The Large Cloud of Magellan is about 160,000 light-years from the sun, the Small Cloud about 200,00 light-years. Hot stars in the two galaxies were exploited as a source of radiation for the study of the galactic corona with the *IUE*. The absorption indicates the presence of gas at temperatures between 10,000 and 200,000 degrees Kelvin. Other data suggest there is also gas at a million degrees K.

formed at a temperature of about 100,000 degrees K., so that their presence gives confirmation for the idea of a hot galactic corona.

In order to gain a more detailed understanding of the absorption lines the television image of the ultraviolet spectrum is converted into a plot of the intensity of radiation at each wavelength. By applying formulas related to the Doppler effect the velocity of the gas responsible for the absorption can readily be computed. If certain additional assumptions are made, the distance of a cloud can be estimated by measuring its velocity. Such a procedure makes it possible to determine the position of hot and cool clouds in the galactic corona.

Since the gas of the galactic disk rotates with the galaxy at 220 kilometers per second, that is the maximum apparent velocity with respect to the sun a cloud of interstellar gas in our galaxy can have. The motion of most points in the galaxy, however, is not directed precisely at the sun. The component of the velocity directed at the sun varies according to the distance from the sun. The sun is itself moving around the center of the galaxy; the apparent velocity of the object is therefore the difference between the velocity of the sun and the velocity of the object along the line of sight. The relation between velocity and distance makes it possible to calculate one quantity from the other.

Along the line of sight to the star designated HD 36402 in the Large Cloud of Magellan the maximum velocity an object in our galaxy can have is 175 kilometers per second away from the sun. When we pointed the telescope of the *International Ultraviolet Explorer* at HD 36402, we observed strong C II absorption at that velocity. We also observed absorption at higher velocities corresponding to gas near the Large Cloud of Magellan. In addition we found evidence of the highly ionized atoms C IV and Si IV at distances perhaps as large as 30,000 light-years (see Figure 6.7). The absorption profile of C IV and Si IV suggests that their density decreases in direct proportion to distance; the density appears to be reduced by a factor of three over 10,000 light-years.

Our results suggest that there are clouds of ionized gas a long way from the galactic plane. The clouds are so far from the plane, in fact, that their

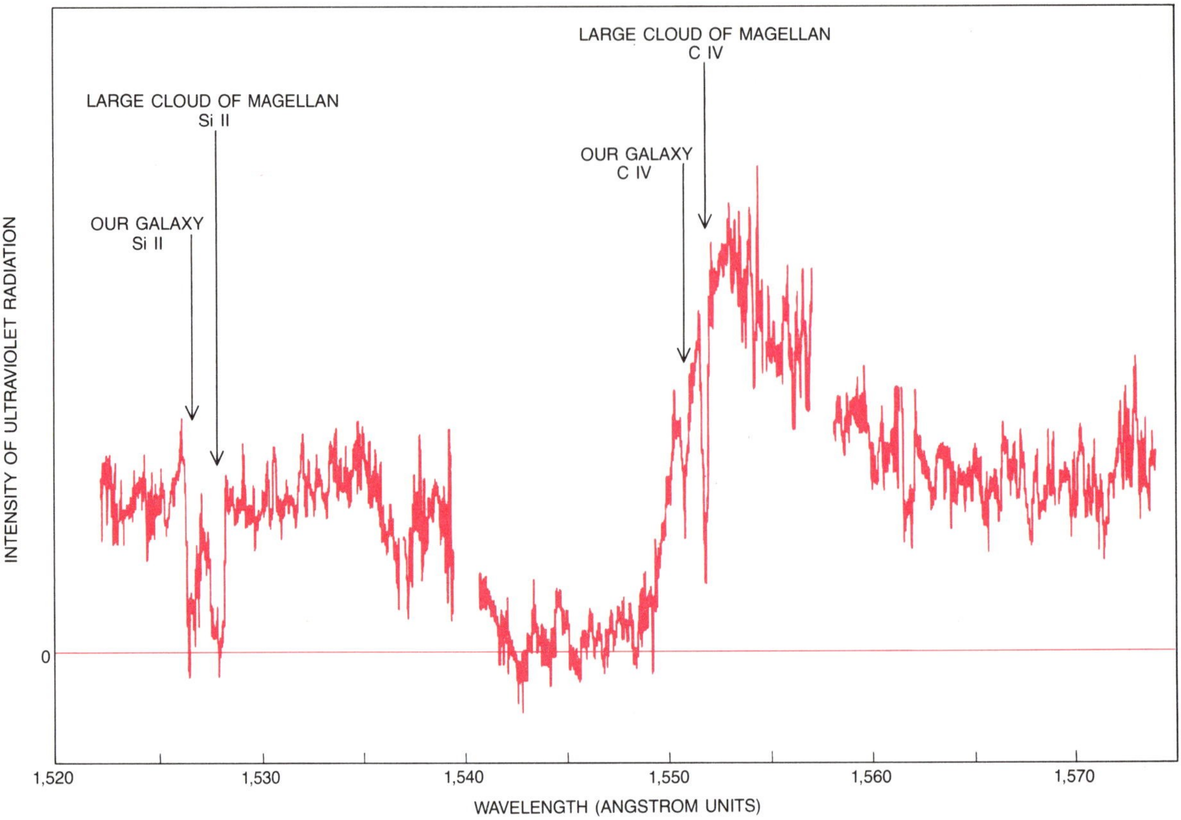

Figure 6.6 ABSORPTION BY THE CORONA of our galaxy and the corona of the Large Cloud of Magellan is shown in a plot of the intensity of radiation detected by the *IUE* against the wavelength. The plot is for three orders of the spectrogram of the star HD 38282 (Figure 6.4); the orders are separated by spaces. Absorption by the coronal gas appears on the plot as deep, narrow troughs. Each ion in the gas absorbs radiation of a particular wavelength. Si II and C IV absorb radiation with wavelengths within the orders shown. The absorption is due to coronal gas and gas in the central plane of the two galaxies. Because of the Doppler effect the wavelength of the absorption by the corona of the Large Cloud is longer than that of the absorption by the corona of our galaxy.

position raises a theoretical problem. The gas of the galactic disk turns at 220 kilometers per second because of the gravitational forces exerted by the stars. The clouds observed with the *IUE* are so far from the galactic plane that one can ask whether the force exerted by the stars is sufficient to keep the gas in the distant clouds moving around the galactic axis of rotation at the same speed as the gas in the galactic disk. If the distant gas does not rotate at the same speed as the disk gas, the calculation of distances from velocities would be invalid.

Other work done with the *IUE*, however, has tended to confirm our estimate of the extent of the coronal gas. G. E. Bromage, A. H. Gabriel and Dennis W. Sciama of the University of Oxford and more recently Max Pettini of the Royal Greenwich Observatory and Kim West of University College London have studied C IV absorption along the line of sight to stars in the galactic halo that are as much as 10,000 light-years away. They found strong C IV absorption in the direction of all stars more than 3,000 light-years from the galactic plane.

The cooler gas of the galactic corona also seems to be widely distributed. With the *IUE* the cool gas is best detected by means of C II absorption. With telescopes on the ground Ca II absorption is the best indicator. Donald G. York and his col-

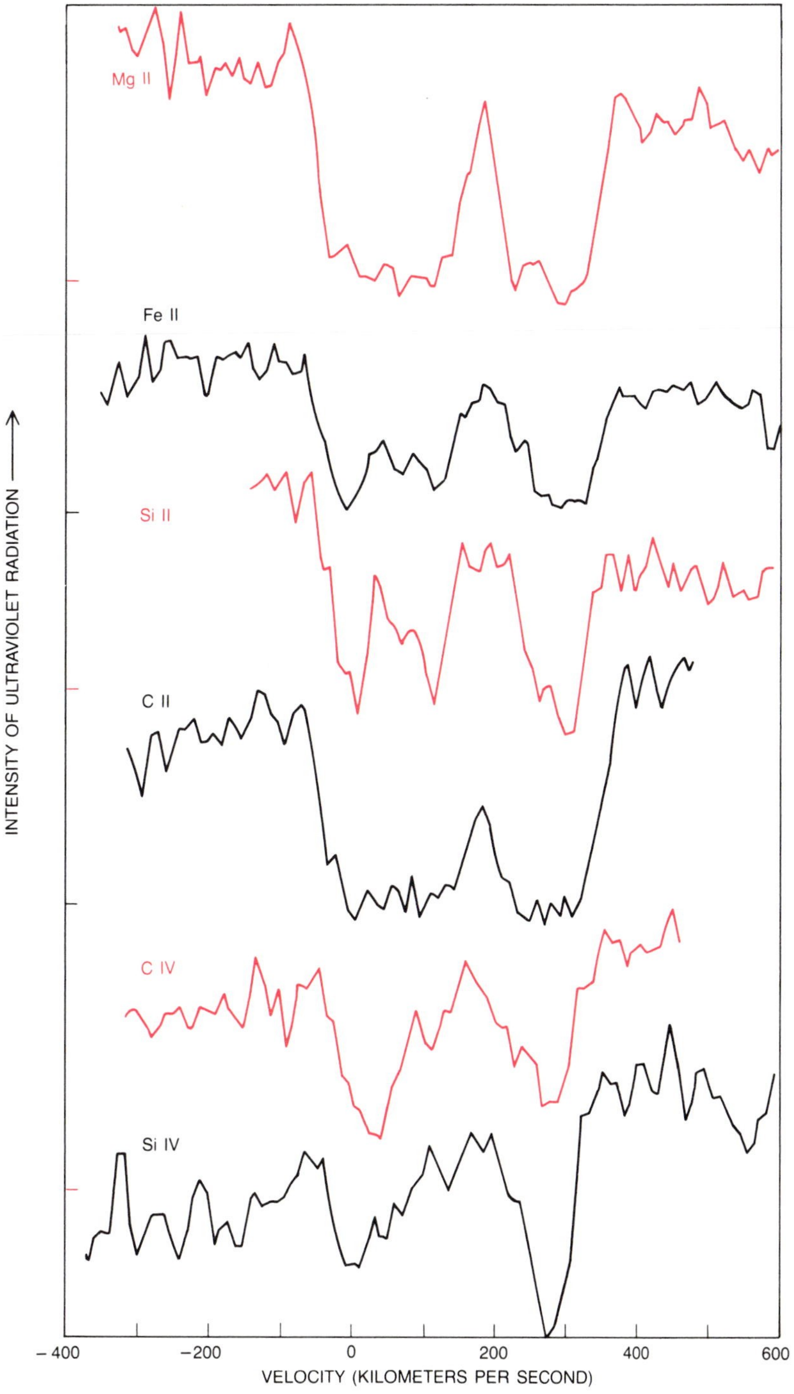

Figure 6.7 VELOCITY OF GAS in the galactic corona can be calculated from the wavelength at which absorption takes place. The intensity of the radiation detected can be plotted against the velocity of the absorbing gas. Shown is a plot for six ions in the coronal gas; the background source was the star HD 36402 in the Large Cloud of Magellan. For each ion strong absorption by stationary gas and by gas with a velocity of 270 kilometers per second is seen; this corresponds to gas near the sun and gas in the Large Cloud. Absorption is also seen at intermediate velocities. If assumptions are made about how the coronal gas moves in our galaxy, the position of the gas can be calculated from its velocity.

leagues at Princeton and Chris Blades of the European Space Agency station of the *IUE*, working with investigators using the Anglo-Australian telescope in New South Wales, have observed Ca II absorption at very large distances from the galactic plane, thereby extending the pioneering work of Münch. Numerous cool clouds, in many different directions from the center of the galaxy, thus seem to be embedded in the hot coronal gas.

Detailed examination of absorption lines at the velocity of the Clouds of Magellan indicates that these two small galaxies also may be surrounded by envelopes of coronal gas much like that of our own galaxy. Our work suggests that all spiral and irregular galaxies have coronas. Further investigation will certainly be needed to confirm this hypothesis. Meanwhile much detail is being added to the general picture we already have of the gas in our galaxy.

That picture has changed considerably over the past 25 years. Until about 10 years ago the interstellar medium was thought to consist of two main components: cool, neutral gas between the stars of the spiral arms and warm, ionized gas between the spiral arms. The detection of O VI absorption and X-ray emission suggested that on the contrary much of interstellar space is filled with highly ionized gas having a temperature of a million degrees K. Gas at such a temperature cools quite slowly; once it has been heated by the explosion of a supernova it remains hot. A subsequent explosion of a supernova would generate shock waves that could travel easily through the hot gas and keep it ionized.

Paul Shapiro and George B. Field of the Center for Astrophysics of the Harvard College Observatory and the Smithsonian Astrophysical Observatory have argued that the hot gas would not be confined to the galactic disk: it would tend to burst out of the disk and flow away from the central plane. The distance it traveled from the disk would depend on its temperature and on the gravitational forces exerted by the stars of the central plane. As the gas flowed outward it would gradually cool and become denser. Denser clouds would begin to descend; ultimately they might be observed as the cool regions of the corona. It is Shapiro and Field who have given the cycle the name galactic fountain (see Figure 6.8).

Other workers have contributed to the effort to understand the processes in the galactic corona. Joel N. Bregman of New York University has extended Spitzer's work by constructing a model for the "percolating" motion of the gas in the fountain. Donald P. Cox of the University of Wisconsin at Madison has analyzed the large-scale structure of the interstellar medium and its energy balance. Results obtained by both workers support the hypothesis that the driving force of the fountain is the explosion of supernovas. If the hypothesis is correct, the vigor of the galactic fountain and the mixing of hot and cool gas in the disk would depend on the frequency of the explosions.

The gas with a temperature of 100,000 degrees K. detected by the *International Ultraviolet Explorer* is far from the hottest gas in the corona. It is difficult to determine whether the gas detected by means of C IV and Si IV absorption is ascending or descending. It is even difficult to tell whether the temperature ascribed to the gas is correct. In dense gases, such as those on the earth, raising the temperature of the gas adds kinetic energy that is quickly spread evenly throughout the atoms by collisions. As a result the temperature of the gas is closely connected to the degree of ionization.

In very diffuse gases such as those of the galactic corona, however, the relation between temperature and ionization may be much weaker. Collisions between atoms and electrons may be so infrequent that changes in temperature are not spread rapidly through the gas. In addition processes such as the absorption of cosmic X rays by the gas might affect the degree of ionization. The presence of particular ions therefore might not indicate the precise temperature of the gas as a whole.

The discovery of the corona of our galaxy has important implications for a long-standing problem concerning the quasars. The emissions of quasars in the ultraviolet region of the spectrum are red-shifted into the visible region. Most workers interested in quasars have contended that the red shifts indicate the quasars are billions of light-years away in an expanding universe. Some workers have suggested that quasars are nearby objects in which peculiar phenomena are taking place.

A careful examination of quasar spectra has shown that in many of the spectra absorption lines can be put in groups that have red shifts smaller than the red shift of the quasar. In 1969 John N. Bahcall of the Institute for Advanced Study and Spitzer suggested that some of these systems of absorption lines might be introduced by the coronas of galaxies along the line of sight to the quasar (see Figure 6.9). With the results from the *IUE* it is possi-

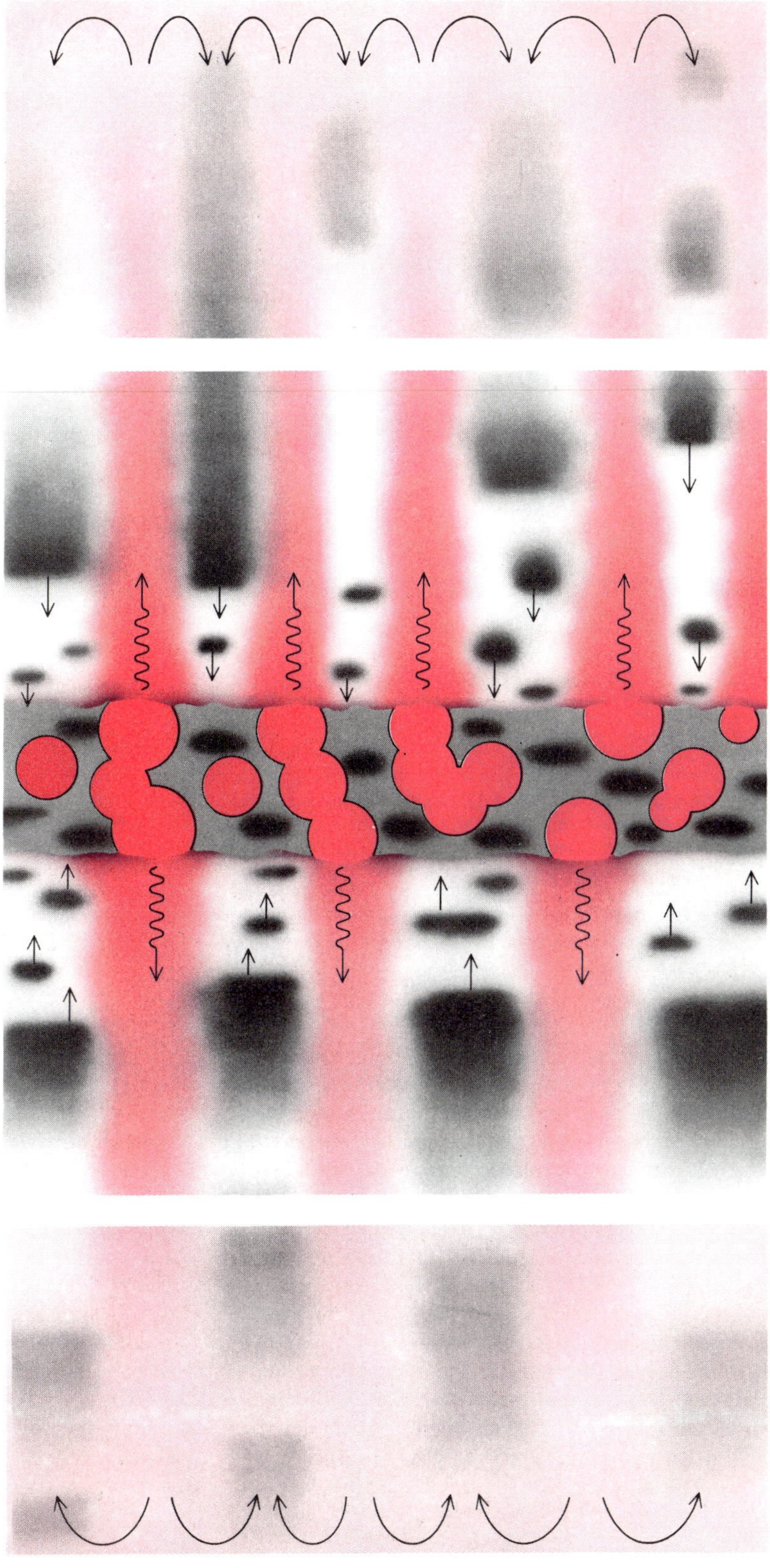

Figure 6.8 GALACTIC FOUNTAIN shown in a schematic cross section of the corona and the galactic disk. The dense gas of the disk is about 500 light-years thick. Hot disk gas rises to a height equal to about 50 times the thickness of the disk. Cooler gas in the form of clouds descends toward the disk. The explosions of supernovas heat some of the disk gas to a temperature of a million degrees K. The heated gas bursts out of the disk and moves away from the central plane. As the gas rises it begins to cool and condense, and then it descends toward the disk. Ultimately it is observed in the form of the clouds of cool coronal gas. When the clouds reach the galactic plane, they become part of the cool gas and the cycle can begin again.

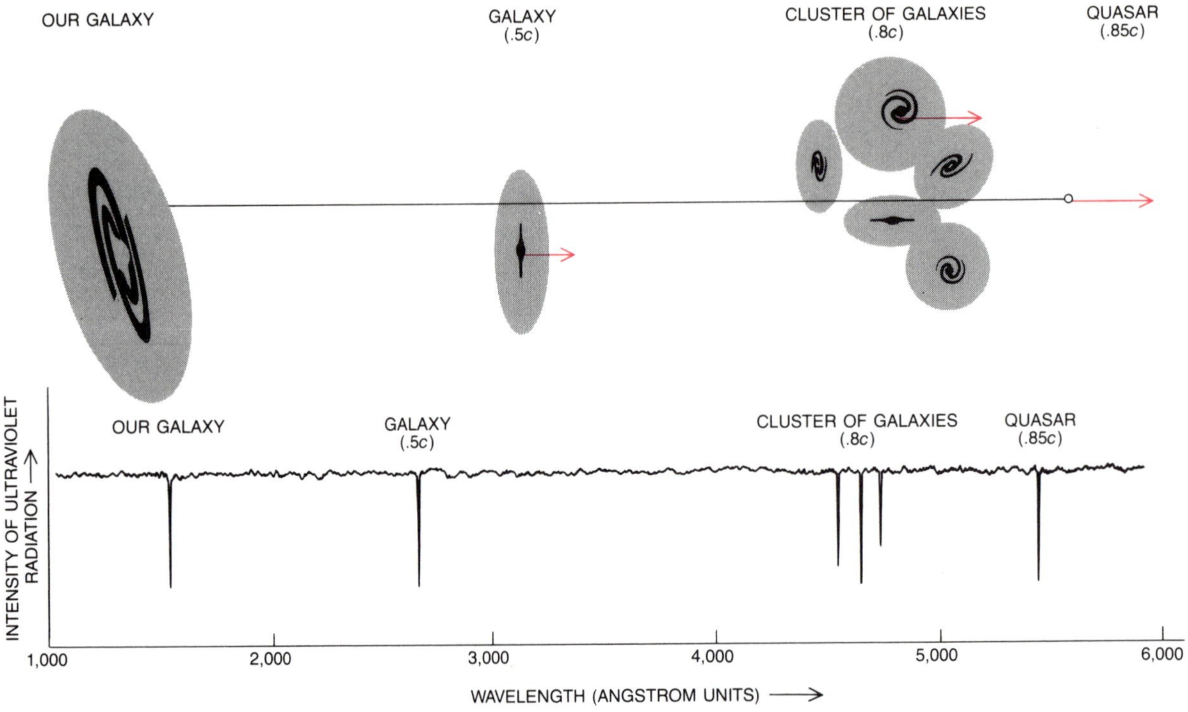

Figure 6.9 SPECTRUM OF A QUASAR (*shown schematically at bottom*) is strongly shifted toward the red, suggesting that the object is moving away from our galaxy at high velocity and therefore is at a great distance. Superposed on such spectra, however, are groups of absorption lines that show smaller red shifts which may be explained by the absorption of radiation from the quasar by the coronas of galaxies between the quasar and our galaxy (*schematic diagram at top*). In the spectrum at the bottom are six absorption lines made by the ion C IV. Stationary C IV absorbs radiation at 1,550 angstroms. C IV absorption near 3,000 angstroms and between 4,000 and 5,000 angstroms corresponds to objects with respective velocities of .5 and .8 the velocity of light (*c*). Absorption near the quasar appears at a velocity .85 that of light.

ble to compare absorption due to the corona of our galaxy with typical quasar absorption-line systems. The patterns of absorption are remarkably similar; the *IUE* data thus support the idea that intervening coronas are responsible for many quasar absorption-line systems.

If the coronas are indeed responsible for the observed absorption, it follows that quasars are at enormous distances and that they represent the most luminous objects known. And if the absorption-line systems commonly seen in the spectra of quasars are due to absorption by the coronas of galaxies, the coronas must be considerably larger than the images of galaxies seen in photographs.

The coronas required to yield the observed effects would need to be from 100,000 to 300,000 light-years in radius, or larger by a factor of three than the dimensions of galaxies in photographs. It has already been recognized, however, that galaxies are much larger when they are observed at the 21-centimeter emission line of neutral hydrogen than when they are observed at visible wavelengths. They would undoubtedly appear larger still if they were probed by means of absorption at the wavelengths of the highly sensitive ultraviolet absorption lines. It should soon be possible to expand greatly on these preliminary studies of galactic coronas. With the launching into orbit of the 2.4-meter Space Telescope much additional information from the ultraviolet part of the spectrum on the corona of our galaxy and the coronas of other galaxies should be forthcoming.

SECTION

III

LUMINOUS STARS

• • •

Epsilon Aurigae

7

Infrared and ultraviolet observations made during the latest eclipse of this binary star system suggest the eclipsing object is a hot, young star surrounded by a large cloud of dust and gas that gave it birth.

• • •

Margherita Hack
October, 1984

Late this month in the middle latitudes of the Northern Hemisphere the constellation Auriga can be seen on the eastern horizon, rising just to the north of Orion shortly after sunset. With the aid of a simple star chart one can quickly find the third-magnitude star labeled epsilon in Auriga, a thumb's width to the southwest of Capella, the brightest star in that region of the sky. In the fall of 1982 a careful observer might have noticed that Epsilon Aurigae was slowly beginning to darken. By December of the same year it had diminished to only half its initial brightness, and it remained a fourth-magnitude star for nearly 11 months. Then gradually it began to brighten until by the middle of last May it had regained its former magnitude.

This cycle of darkening and brightening has been repeated four times in the 20th century, at precise intervals of 27.1 years. The basic mechanism of the cycle was first explained by the German astronomer Hans Ludendorff in 1904. It is the result of an eclipse of the primary, or bright, visible star, by an unseen companion object; the visible star and its companion form a binary system held together by gravity. The length of the cycle is the longest orbital period known among eclipsing binaries, but it is the dark phase of the cycle that is hardest to explain. Its duration implies the eclipsing object must be enormous: the width of the object needed to account for nearly two years of eclipse is about 1,500 times the radius of the sun. Although each eclipse has been studied by a new generation of astronomers with new and more precise observational techniques and with increasingly sophisticated theory, the identity of the eclipsing object remains the great mystery of Epsilon Aurigae.

The eclipse of 1982–84 was the first one in which the star could be observed from satellite platforms, and the results have extended observational knowledge of Epsilon Aurigae into the far ultraviolet. At the other end of the electromagnetic spectrum, observations have been made in the deep infrared by instruments such as the three-meter Infrared Telescope Facility of the National Aeronautics and Space Administration at the summit of Mauna Kea on the island of Hawaii. The data gathered during the most recent eclipse are still being studied and were the topic of a workshop held in Tucson, Ariz., in January, 1985. They confirm earlier evidence that Epsilon Aurigae is a unique object in the Milky Way, and they have already made it possible to eliminate several earlier models that were proposed to account for the observations (see Figure 7.1).

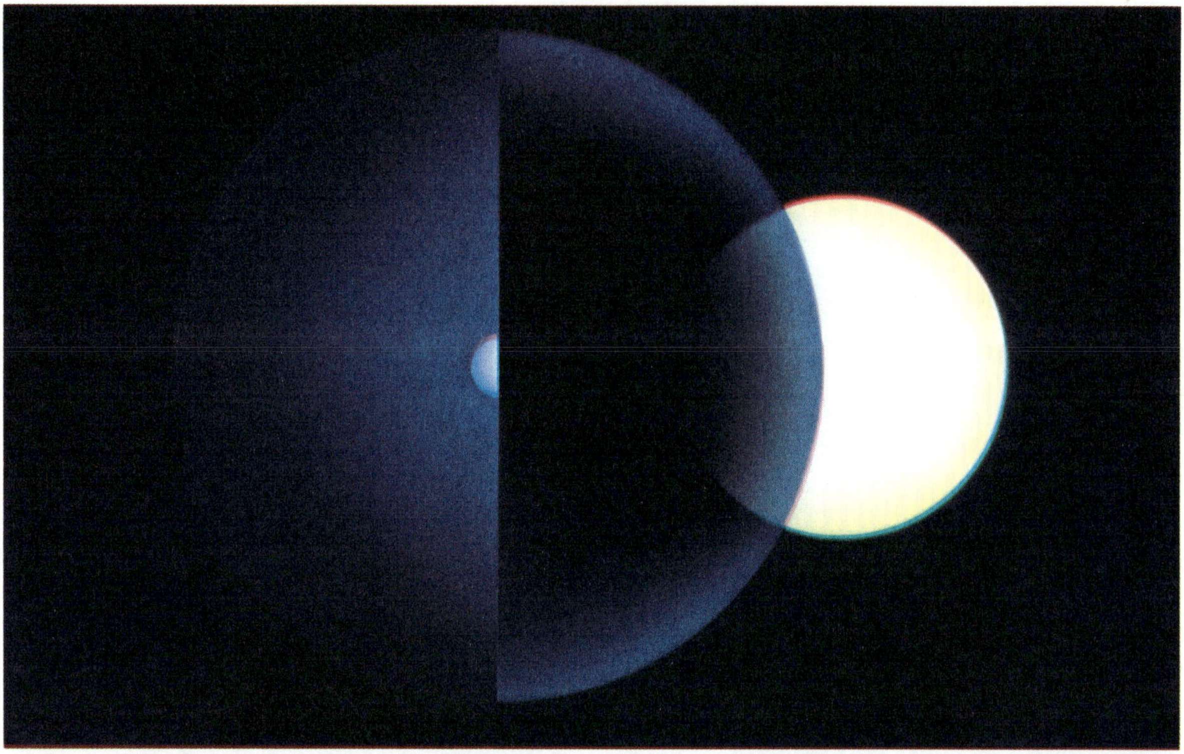

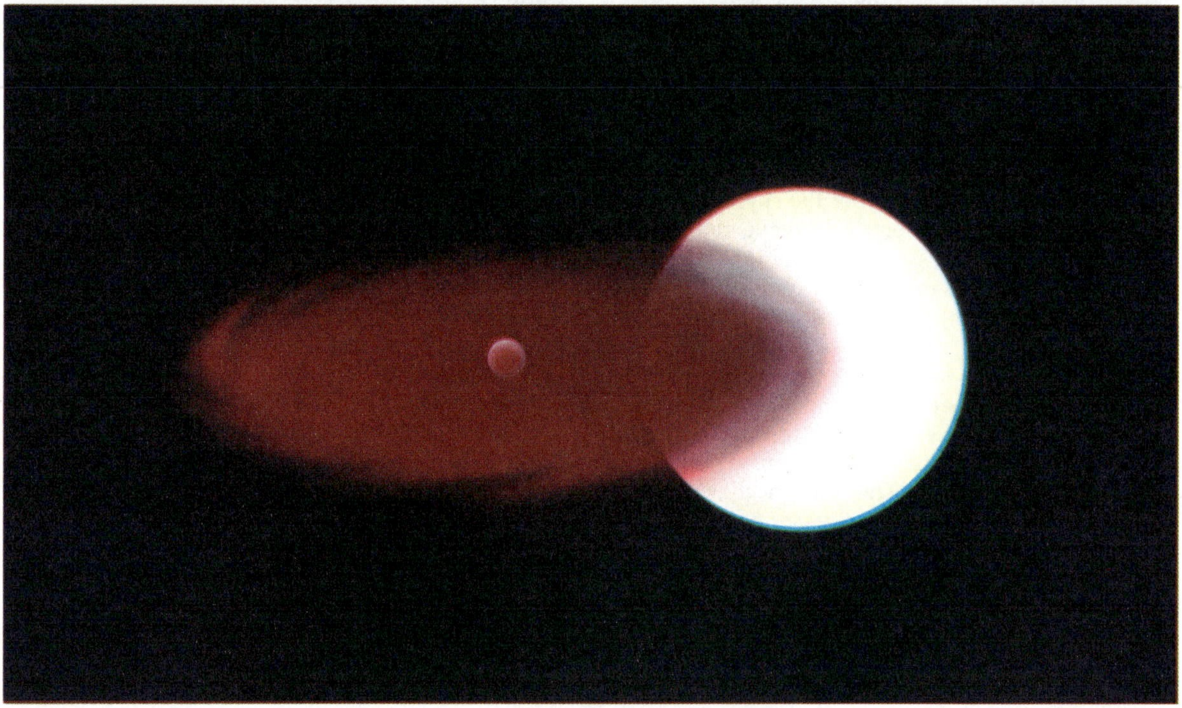

Nevertheless, by building on and combining some of the earlier models of the star one can now give a plausible and rather detailed interpretation of the spectroscopic and photometric code carried to us from a star 1,900 light-years away.

The visual observation of an eclipsing binary system is not unusual by itself. At least half of the stars in the Milky Way are found in gravitationally bound binary systems, and many of these systems are oriented in such a way that from our vantage in space stars periodically eclipse each other. All eclipsing binary systems are too distant to be resolved into their stellar components, and so from our vantage such a system appears to be a single star. Nevertheless, an eclipse of the system can cause the apparent star to change its brightness periodically in much the same way the headlight of an oncoming car flickers briefly in the night when its path its crossed by an otherwise invisible pedestrian.

The two-year-long dimming of the light of Epsilon Aurigae was the first evidence that the star is unusual, but the spectroscopic data suggest the true dimensions of the problem. By measuring the Doppler shift in the spectral lines of the primary star one can determine the component of its velocity along the line of sight between the star and the earth. That component enables one to estimate the orbit of the primary star about its center of gravity and to give upper and lower limits for the mass of the companion. The orbital characteristics of the system suggest the mass of the companion is between four and 15 times the mass of the sun.

Given such a mass and the enormous size implied by the length of the eclipse, it is surprising the companion is not visible even with the largest telescopes. During the secondary eclipse of the companion star by the primary star it should be possible to detect a companion star up to three magnitudes fainter than the primary star, or a factor of about 16 in brightness. That sensitivity, however, can be attained only if the orbital plane of the system is nearly parallel to the line of sight of the observer; if the inclination of the system is such that the eclipsing primary star only grazes the disk of the companion, a brighter companion star could go undetected. Nevertheless, in observations with ground-based instruments made before and after eclipse, the spectrum of the companion star is completely overwhelmed by the spectrum of the primary star. The companion must therefore be at least two magnitudes darker than the primary star.

Figure 7.1 AN EARLY MODEL of Epsilon Aurigae suggested the eclipse is caused by a shell of ionized gas and free electrons that surrounds a hot companion star, which is observable only in the ultraviolet (*upper illustration*). A quarter sector of the shell has been cut away to reveal half of the disk of the companion. Another early model suggested the eclipsing body is a disk made up of grains of dust whose average diameter is significantly larger than the wavelength of infrared radiation (*lower illustration*). Recent observations in the ultraviolet and infrared regions of the spectrum indicate that neither model can entirely account for the data.

One might suppose that during the eclipse of the primary star by the companion the spectrum of the companion could be discerned. What is seen instead is one of the most puzzling aspects of the Epsilon Aurigae system and one of its most telling features. The lines in the spectrum of the system during the eclipse are almost identical with the lines seen before and after the eclipse. The observation might suggest both objects in the system belong to the same spectral class and have the same surface temperature, but that explanation is ruled out by the failure to observe the companion before and after the eclipse.

The alternative explanation is simply that the spectrum of the primary star must always be visible. For example, the disk of the primary might not be completely occluded by the companion, or it might be eclipsed by a cloud of dust or gas that allows some of the light of the primary star to pass through. Whatever model is adopted, it must also be reconciled with the observation that the intensity of the light emitted by the primary star is reduced almost uniformly during the eclipse, regardless of the wavelength at which the event is observed (see Figure 7.2).

What can account for the unique set of observations? Ludendorff suggested that the companion is not a star at all but instead is a swarm of meteoroids in orbit about the primary star. That suggestion was revived in the 1960's in somewhat revised form, but at the time of the initial suggestion the huge mass that would have been assigned to the meteoroidal cloud would have been unacceptable to most astronomers.

In 1937 G. P. Kuiper, Otto Struve and Bengt Strömgren of the Yerkes Observatory proposed a model based on observations of the eclipse of 1928–30. According to their model, the companion

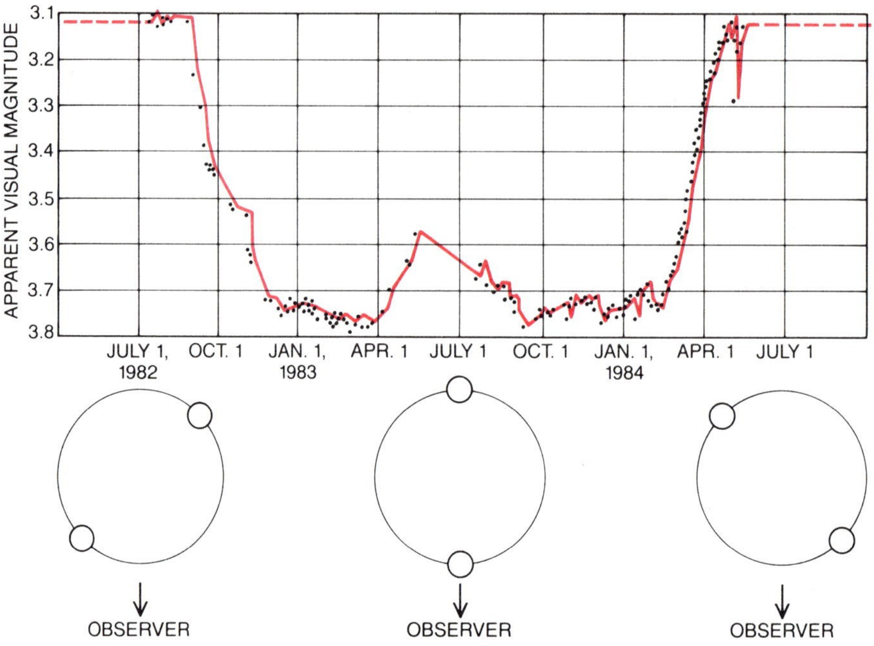

Figure 7.2 LIGHT CURVE in which the intensity of the light emitted by Epsilon Aurigae is plotted against time. Below the curve the eclipse is depicted schematically: part of the light from the more distant of the two stars in the system is periodically obscured by an object along the line of sight between the distant star and the earth. The light curve shows that the total light output of Epsilon Aurigae decreases during eclipse by a magnitude of .8, or about half its brightness out of eclipse. The partial phases of the eclipse, both ingress and egress, last for 192 days each, and totality, or the maximum depth of the eclipse, lasts for 330 days. The data were plotted by observers at the Hopkins Phoenix Observatory and the Tjornisland Astronomical Observatory.

is an enormous but extremely tenuous mass of gas called an *I* star, which is partially transparent to light. The bulk of gas needed to account for both the probable mass of the companion and the duration of the eclipse must be so rarefied that a mechanism had to be introduced to increase the opacity of the eclipsing *I* star.

The mechanism is analogous to the one responsible for the formation of the earth's ionosphere. The atoms in the outer layers of the star were to be ionized, or stripped of one or more of their electrons, by ultraviolet radiation emitted by the primary star. Free electrons absorb light and scatter it in a way that is effectively independent of wavelength. Hence the emissions of the primary as seen from our solar system could be uniformly attenuated by passing through the ionization layer of the *I* star. The temperature of the *I* star was assumed to be about 700 degrees Celsius, which is far too cool for detectable emissions of visible radiation and so accounts for the failure to observe the companion star.

One of the main criticisms of the 1937 model was that even the free electrons would not account for the reduction in the light output during the eclipse. In 1954 Zdeněk Kopal of the University of Manchester proposed that instead the eclipsing body is a ring of solid particles or dust grains surrounding the invisible companion star. The ring is assumed to be thin enough so that when viewed roughly edge on it obscures only half of the visible hemisphere of the primary star. Kopal has since suggested the disk closely resembles a protoplanetary nebula, a rotating disk of gas and dust from which a solar system can be formed. In this model too the spectrum of the primary star is always observable, and the eclipsing body, which only absorbs the light of the primary, does not contribute appreciably to the light of the system.

Almost all astronomers agree with the basic geometric picture described by the last two models. In other words, the eclipsing body is either a semitransparent shell absorbing half of the light of the primary or it is a flat disk that covers only half of the hemisphere of the primary. There is almost no agreement, however, on the answers to two major questions that appear capable of being resolved: What is the composition of the eclipsing matter, and what is the nature of the invisible companion?

Only two kinds of material are known that can absorb light independently of wavelength. I have already mentioned that a gas of free electrons can absorb and scatter all wavelengths equally. The second candidate material is a cloud of dust grains. If the diameter of each grain is substantially larger than the wavelength of infrared and visible radiation, a cloud of dust would absorb radiation of all detectable wavelengths. According to several models of Epsilon Aurigae, which collectively have both shell and disk geometry, dust is the main cause of the eclipse.

A close study of spectrums made both in and out of eclipse, however, tends to contradict the dust-grain hypothesis. The spectrums give a subtle but clear way to distinguish the light scattered by the eclipsing body from the light that passes unimpeded from the primary star to the earth. The spectrum of the primary star is made up of a series of dark absorption lines set against a background of emitted light that forms a continuum of spectral colors. The absorption lines are caused by the excitation of atoms in the photosphere, or outer and visible layer, of the star. According to the laws of quantum mechanics, the atoms can be excited only to certain discrete energy levels, and so the energy required to excite them must take on discrete values corresponding to the allowed excitation levels. Photons that carry the discrete values of energy are thereby absorbed and the rest of the energy is transmitted; the result is an absorption spectrum characteristic of the atoms that make up the photosphere.

If some of the light emitted by the star were to pass through a shell or a disk made up of dust, the light would be scattered in all directions. The spectrum of scattered light would be identical with the spectrum of the primary star observed outside the eclipse, except for a small correction. The stars in a binary system generally rotate about their axes in the same clockwise or counterclockwise sense as they revolve about their common center of mass. If the eclipsing body is a shell or a disk, it too must rotate, and during the ingress phase of the eclipse, just before totality, the part of the shell or disk that eclipses the primary rotates away from the observer (see Figure 7.3). The scattered absorption lines are thereby shifted toward the red end of the spectrum by the Doppler effect.

During the egress phase of the eclipse, just after totality, the effect is reversed. The part of the shell or disk that eclipses the primary rotates toward the observer, and the scattered light is Doppler-shifted toward the violet. During totality the shift is zero because it is the central part of the shell or disk that eclipses the primary; the radial component of its rotational velocity is zero. The observed spectrum is the superposition of two components, namely the spectrum of scattered light and the spectrum of light that reaches the earth directly from the primary star without being scattered. Thus just before and just after totality the absorption lines of the primary star would be broadened slightly in the direction of the Doppler shift and the difference between the light intensity in the darkest part of the line and the light intensity of the surrounding continuum would diminish.

If the light from the primary star passes instead through a gas of free electrons, the observed spectrum is slightly more complex. Not only is the light scattered by the electrons but also it is transformed by its passage through the gas of ionized atoms from which the free electrons are derived. The gas and the photosphere of the primary star have approximately the same atomic composition, and the gas too absorbs photons of light whose wavelength corresponds to some excited atomic-energy level. The absorption lines in the resulting spectrum are again added by superposition to the lines in the spectrum of any light that reaches the earth directly from the primary star.

Because of the additional absorption by the gas, however, the absorption lines are enhanced during totality, or darkened with respect to the background continuum. During ingress and egress the absorption lines from the shell or disk of gas are Doppler-shifted, just as they are when the light passes through a cloud of dust. The enhancement of a line combined with the Doppler shift often splits the line into two sharp points. The double-pointed lines in the absorption spectrum provide an unmistakable signature of the shell or disk that scatters the light; the shifted component of the lines is called the shell spectrum.

Doppler-shifted lines in the spectrum of Epsilon Aurigae were noted by Ludendorff in 1901 and became widely known after the eclipse of 1928–30. The lines show a clear enhancement attributable to the shell spectrum, and many of them have two sharp points during ingress and egress. Consequently there seemed little doubt that the eclipse of the primary star is caused by gas rather than by dust.

The record of the eclipse of 1928–30 confirmed this conclusion in a surprising way. Struve noted that certain absorption lines, which are formed only when atoms are excited to a high energy level, and are not enhanced during totality and are not split in two during the partial phases of the eclipse. For example, during eclipse there is no detectable shell component to the blue line with a wavelength of 4,481 angstrom units that is associated with singly ionized magnesium. (One angstrom unit is 10^{-10} meter.) If the eclipsing body were made up entirely of dust, the blue magnesium line would be Doppler-shifted by the shell or disk, and the change in the shape of the line would be observable in the spectrum. On the other hand, if a gas of atoms and free electrons eclipses the primary star, the blue magnesium line is easy to understand. The temperature and the density of the shell or disk of gas are presumably much lower than the temperature and the density of the stellar photosphere where the line is initially generated. Thus the number of photons and particles in the shell or disk energetic enough to cause atomic excitation at high energies is too small to bring about a measurable enhancement in the corresponding absorption lines.

In 1954 Robert P. Kraft, then a young associate of Struve's at the University of California, Berkeley, found a difficulty with the electron-scattering hypothesis. Struve had assumed the primary star was the source of the energy needed to ionize the atoms in the eclipsing shell or disk of gas and thereby create the free electrons. The energy output of the primary star, however, is known by direct observation. Its surface temperature is about 7,500 degrees C. and its radius is about 100 times the radius of the sun. Kraft calculated that the electron density in the shell or disk that could be generated by such a star would be about 100 million electrons per cubic centimeter. That density is too small, by a factor of about 1,000, to explain the depth of the eclipse. In other words, if the opacity of the shell or disk is attributed to scattering by electrons generated by the primary star, the shell or disk is still too transparent to explain the observations.

Kraft suggested the opacity of the gas is caused instead by the presence of negative hydrogen. Negative hydrogen is an ordinary hydrogen atom that has captured an extra electron. It can form from the excess electrons generated in a gas hot enough to ionize atoms such as iron, manganese and chromium but not hot enough to ionize neutral hydrogen. In the limited spectral range observable from the ground the absorption of radiation by negative hydrogen is almost independent of wavelength; consequently when Kraft introduced his idea it seemed quite promising. In the ultraviolet, however, the absorption of radiation by negative hydrogen becomes much smaller, and so it would be too transparent to account for more recent observations of Epsilon Aurigae made at ultraviolet wavelengths high above the atmosphere.

I began my own observations of Epsilon Aurigae during the eclipse of 1955–57, and in 1957 I had the opportunity to study the spectrums made by Struve during that eclipse. The resolution of the new spectrums was much higher than it was for the spectrums made during the previous eclipse. I was thus able to demonstrate the presence of certain weak shell lines such as neutral, excited magnesium and calcium that had formerly seemed to be missing. By measuring the intensity of the newly detected lines and by counting the number of lines in the hydrogen spectrum I was able to make a direct estimate of the electron density in the shell or disk of gas, without regard for the source of ionizing energy. The density turned out to be about 100 billion electrons per cubic centimeter, or 1,000 times the density calculated by Kraft. Given a shell or disk with a thickness of about .7 astronomical unit, such a density would be just sufficient to explain the observed depth of the eclipse by electron scattering alone. (An astronomical unit is the mean distance between the earth and the sun, or about 93 million miles.)

Since Kraft had shown the primary star is not hot enough to produce the necessary electron density, I still faced a major problem: What is the source of energy that ionizes the shell or disk? In 1962 I suggested the ionization is caused by the invisible companion star, which is surrounded by a thick shell of gas. The shell of gas is dragged along by the com-

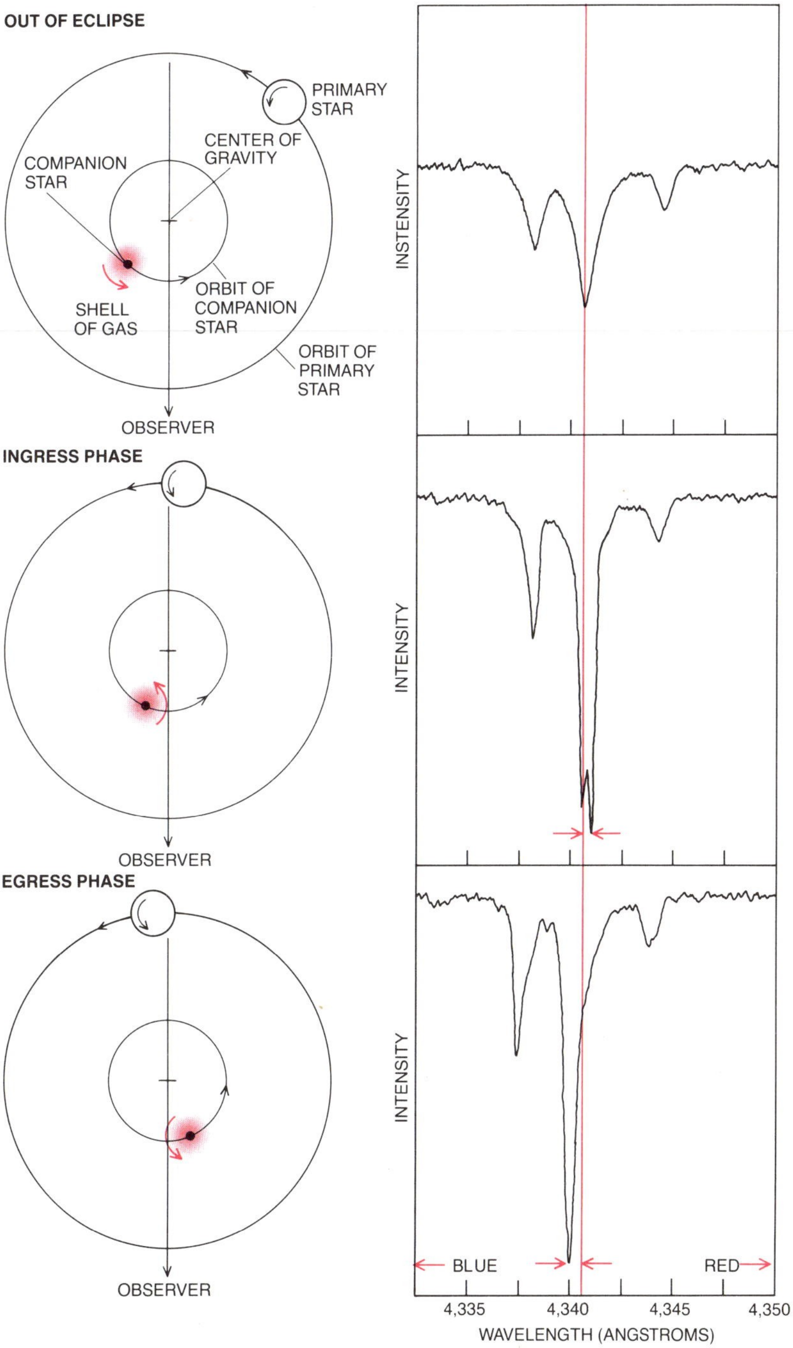

Figure 7.3 "SHELL SPECTRUM" of Epsilon Aurigae. At top the blue region of the spectrum is plotted near one of the Balmer lines of hydrogen at 4,340 angstroms. During the ingress and egress phases the line deepens and splits into two sharply defined regions of minimum intensity. During the ingress phase the rotation of the cloud is such that the gas along the line of sight between the primary star and the earth is moving away from the observer, and so the spectral absorption line is shifted toward the red (*middle*). During the egress phase the absorption line is shifted toward the blue because the eclipsing part of the cloud is moving toward the observer (*bottom*). (Spectrums by the author and collaborators with the coude spectrograph of the 152-centimeter telescope of the Observatoire de Haute Provence.)

panion as it moves in its orbit, and the eclipse is observed when the shell is imposed between the earth and the primary star. I was able to calculate a range of possible temperatures and radii for the invisible companion from the characteristics of the shell spectrum. The calculations showed the companion could be a giant star with a surface temperature of 15,000 degrees C. and a radius 60 times that of the sun, or it could be a hot subdwarf with a temperature of about 100,000 degrees and a radius about equal to that of the sun.

At the time I made my proposal the detection of a hot companion star seemed unlikely. It is well known that hot stars radiate most of their energy in the ultraviolet region of the spectrum, and the radiation is almost completely absorbed by the earth's atmosphere. Confirmation of a hot star in Epsilon Aurigae had to await observations at ultraviolet wavelengths between 1,000 and 3,000 angstroms made from astronomical satellites.

The first attempts to observe such a star were made with a telescope built at Princeton University and mounted on the NASA *Copernicus* satellite, and with a spectrometer designed by astronomers from Belgium and Great Britain and mounted on the *TD-1* satellite of the European Space Agency. The Princeton telescope, built to obtain high-resolution spectrums of hot stars brighter than the fifth magnitude, gave no positive evidence for the hot star. The Belgian and British spectrometer, built to obtain low-resolution spectrums of hot stars brighter than the seventh magnitude, gave a few uncertain results. There was some evidence for ultraviolet radiation from Epsilon Aurigae in excess of that expected from the primary star alone.

In January, 1978, the *International Ultraviolet Explorer* satellite (*IUE*) was launched, and after a few months of testing in orbit the satellite was made available to guest observers. The *IUE*, a joint venture of NASA and the European Space Agency, was designed to give high-resolution spectrums of stars as faint as the 10th magnitude, as well as low-resolution spectrums to the 13th magnitude. On April 19, 1978, my colleague Pier Luigi Selvelli of the *IUE* European observing station near Madrid and I observed the low-resolution spectrum of Epsilon Aurigae on the screen of a video terminal. There for the first time was the spectral signature of the binary system in the far ultraviolet, at wavelengths as short as 1,300 angstroms. The presence of the previously undetected companion star was unambiguously revealed.

We have since made several additional observations out of eclipse, and two of my colleagues at the Trieste Astronomical Observatory, Conrad Boehm and Steno Ferluga, and I have also had the opportunity to observe Epsilon Aurigae in the ultraviolet during the most recent eclipse. The observations confirm the existence of a hot companion whose light is dominant at wavelengths shorter than 1,550 angstroms. More precisely, the dimming of the light from the system during eclipse is between .8 and 1 magnitude for all visible and ultraviolet wavelengths as short as 1,600 angstroms, whereas almost no trace of the eclipse is observable in the far ultraviolet, between 1,240 and 1,550 angstroms. Because the flux of the relatively cool primary star is negligible in the far ultraviolet, what we see in that spectral range is the flux of the companion star, which is in the foreground during eclipse (see Figures 7.4 and 7.5).

Although the new observations have been dramatic, they have not given detailed confirmation of the model I proposed in 1961. The ultraviolet spectrum of the companion star shows it is not as hot as I had predicted it to be. Its surface temperature is about 10,000 degrees C. and its radius is between three and five times the radius of the sun.

Moreover, the infrared spectrum appears to be in complete contradiction with the data in the ultraviolet. Observations in the infrared made by Dana E. Backman, Eric E. Becklin, Dale P. Cruikshank, Theodore Simon and Alan T. Tokunaga of the University of Hawaii at Manoa and by Richard R. Joyce of the Kitt Peak National Observatory have monitored Epsilon Aurigae before and during the eclipse at wavelengths as long as 20 micrometers. The investigators have found the eclipse dims the light by about .7 magnitude at wavelengths between one micrometer and 4.8 micrometers, which is roughly equal to the depth of the eclipse for visible light. At longer wavelengths, however, the depth of the eclipse becomes smaller, and at a wavelength of 20 micrometers it is only .3 magnitude, down by a factor of about 1.3 in brightness (see Figure 7.6). From these observations Backman and his colleagues concluded the eclipsing body is a cool object whose light is dominant in the far infrared. They derive a surface temperature of about 200 degrees C. and a radius of 10 astronomical units.

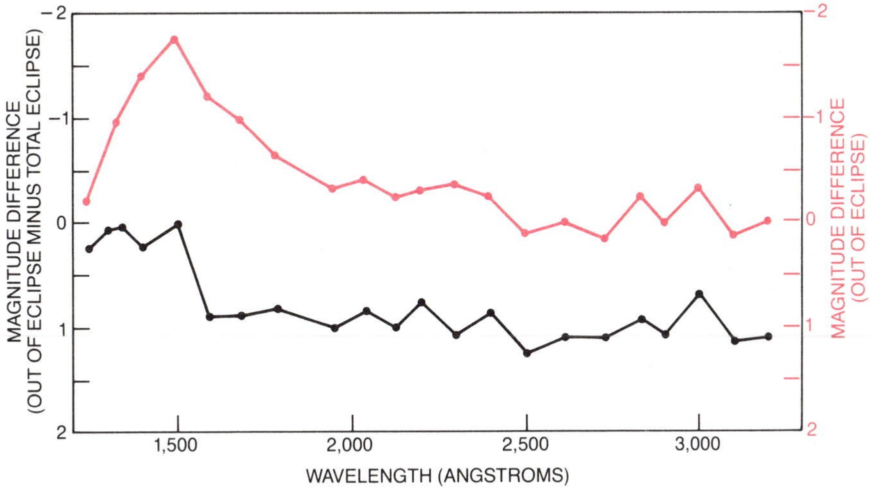

Figure 7.4 DIFFERENCE BETWEEN THE MAGNITUDE of Epsilon Aurigae before eclipse and its magnitude during eclipse is plotted in black. In the range of near-ultraviolet wavelengths between 1,600 and 3,000 angstrom units the difference oscillates between .8 and 1 magnitude. In the far ultraviolet, however, at wavelengths shorter than 1,600 angstroms, the difference is less than .2 magnitude. What is seen is the light of the hot companion star, which is in the foreground during the eclipse and emits radiation primarily in the far ultraviolet. Also plotted against wavelength is the magnitude difference between observations made at two different times before the eclipse (*color*).

The infrared observations also pose a more subtle problem for my 1961 model. In a completely ionized shell of gas there is a process called free-free absorption that leads to strong emissions of radiation at infrared wavelengths. When a free electron in the gas passes near a proton or a positive ion, the electron readily gives up some of its energy in the form of an infrared photon; the greater the number of electrons and the more energy that has been absorbed by the gas from an external source, the more likely the process. The observed flux of infrared radiation could be interpreted as partial confirmation of the presence of free-free absorption in a gas surrounding the companion star. The flux, however, is much smaller than it would be if it were caused by free-free absorption in a shell as thick as the one in my early model.

How can one construct a model that explains such varied and apparently conflicting observations? It is relatively straightforward to adjust the temperature of the companion star in my early model to match the observed temperature. Remember that the high temperatures in my model were necessary to account for the energy needed to ionize the shell. In the early 1960's stellar radiation was thought to be the predominant and perhaps the only ionizing agent. Since that time, however, observations in ultraviolet regions in the spectrums of several stars have repeatedly shown the presence of lines from multiply ionized elements, which cannot be generated solely by the radiation in the stellar photosphere. The phenomenon is called superionization, and it strongly suggests that other effects, of mechanical or magnetic origin, can contribute to the ionization of the gas. If superionization is the cause of the ionization of the eclipsing shell in Epsilon Aurigae, the relatively low, 10,000-degree temperature of the companion star could still give rise to the opacity of the ionized shell.

The observations in the infrared, however, suggest the need for more substantial changes in the model. The infrared radiation generated in my model by free-free absorption can be reduced to match the observed infrared flux if the thickness postulated for the shell of gas is reduced as well. In that case, however, the shell of gas alone cannot account for the observed opacity of the eclipse. The promising solution to the dilemma appears to be a model that combines aspects of the two most popu-

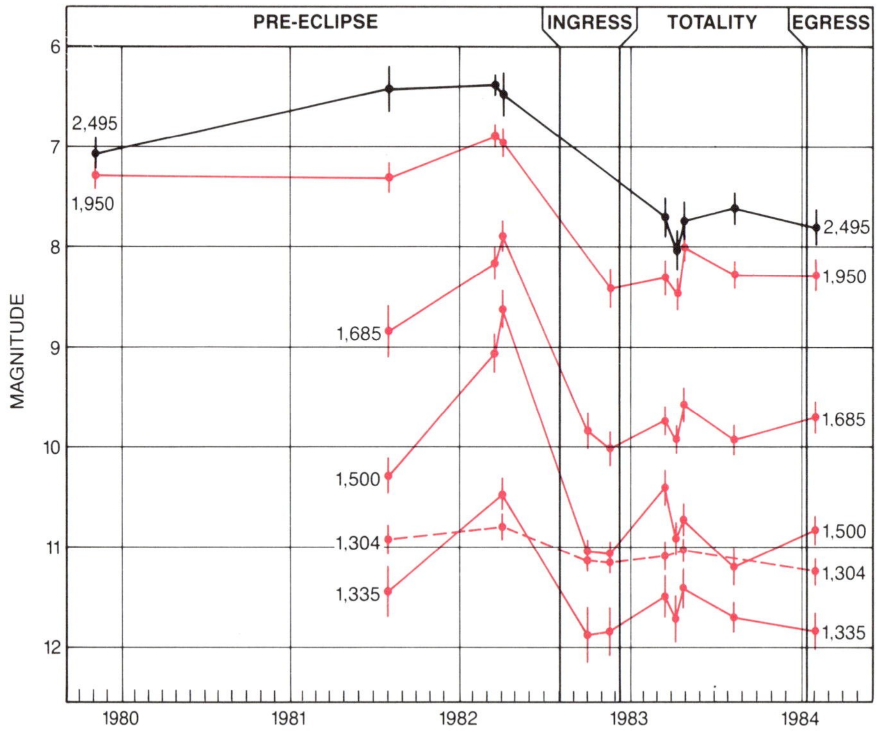

Figure 7.5 ULTRAVIOLET LIGHT CURVES give the radiation variations from Epsilon Aurigae. At 2,495 angstroms in the near-ultraviolet region of the spectrum (*solid lines in black*) the intensity is roughly constant before the eclipse and falls smoothly through the ingress phase to a minimum during totality. At almost all wavelengths shorter than about 1,950 angstroms, however, the intensity of the light is greater at certain times during the totality of the eclipse than it is during the ingress phase (*solid lines in color*). The intensity of the emission line of neutral oxygen at 1,304 angstroms (*broken lines in color*) is approximately constant both in and out of eclipse. (Data obtained by Conrad Boehm, Steno Ferluga and the author from the *International Ultraviolet Explorer* satellite.)

lar models of the 1950's: the one that postulates a disk of dust and my model, which postulates the shell of gas.

In the new model the main cause of the eclipse is a ring of large dust particles, which absorbs half of the light of the primary star. The hot companion star is embedded in the ring and appears less luminous than it is because the ring absorbs part of its light. The ring is heated slightly by the companion star and emits some of the observed infrared flux as thermal radiation. To account for the shell spectrum, as well as for the fact that the appearance of the shell spectrum slightly precedes the absorption of light at the beginning of the eclipse, the model postulates a shell made up of gas that envelops the ring of dust. The gaseous shell is ionized by the radiation of the hot star emitted in a direction perpendicular to the plane of the ring, and it is formed of matter that could be either escaping from the hot star or transferred to it by the primary star. The shell of gas is not as thick in the new model as it is in my earlier model, but it extends to a distance of about 10 astronomical units from the hot star (see Figure 7.7).

At least two additional observational features of Epsilon Aurigae can be incorporated in the new model. An extended shell of ionized gas should give rise to an emission spectrum out of eclipse as well as the observed absorption spectrum during the eclipse. The emission lines are not observed in the visible spectrum, probably because they are fainter than the continuum emission of the primary star. A few emission lines, however, are observable in the ultraviolet, namely neutral oxygen at a wavelength

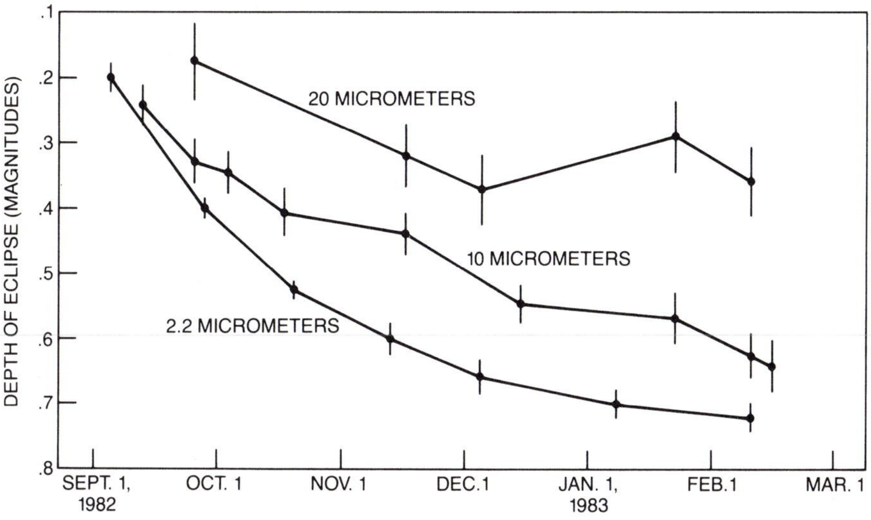

Figure 7.6 INFRARED LIGHT CURVES for emissions of Epsilon Aurigae in the near infrared, such as the one at a wavelength of 2.2 micrometers, exhibit roughly the same reduction in magnitude during the eclipse as the visible light curves do. At longer wavelengths, however, the effect is much smaller; at 20 micrometers, for example, the intensity of the emission is reduced by only about .3 magnitude, or a factor of 1.3 in brightness. The data initially seemed to contradict the interpretation of the ultraviolet light curves: they suggested the eclipsing object in Epsilon Aurigae is relatively cool instead of being hot.

of 1,304 angstroms and singly ionized magnesium at wavelengths of 2,795 and 2,802 angstroms. The intensity of the lines is roughly constant both during and out of eclipse. Such lines must be formed either in a part of the eclipsing shell of gas that is not intercepting the light of the primary star or in an extended shell of gas that surrounds the entire binary system.

The second feature may be a clue to the evolutionary stage of the Epsilon Aurigae system. Observations in the ultraviolet show the companion is a variable star; in the far ultraviolet the variability is between magnitude 1 and 1.5, or in other words a factor of between 2.5 and 4 in brightness. The variability is similar to a rare group of stars in the same spectral category as the companion called Herbig variables, after George H. Herbig of the University of California, Santa Cruz. The Herbig variables are always found in young stellar clusters that show large and irregular variations in their emitted light. The stars themselves are thought to be still contracting from a cloud of gas and dust. Indeed, many Herbig variables are surrounded by clouds of dust whose dimensions and temperature are similar to the object postulated by Backman and his colleagues to account for the infrared observations of Epsilon Aurigae.

If the companion star in Epsilon Aurigae is a young star, one can readily account for the fact that no similar binary system has been observed. The closest relatives known are other long-period eclipsing binaries such as 31 Cygni, 32 Cygni, Zeta Aurigae and VV Cephei. In all four systems a relatively hot star is gravitationally bound to a cool giant or supergiant star. There is no evidence, however, that the hot star in any of these systems is surrounded by such an extended shell of dust as the one observed in Epsilon Aurigae. The difference, I propose, is that Epsilon Aurigae is a similar system in a different stage of its life history.

The primary star in Epsilon Aurigae is a supergiant, but it is hotter than the supergiants in the other four binary systems I have mentioned. Its surface temperature and spectral class show it is undergoing rapid evolution, and it may have recently—that is, in the past million years—completed a phase of abundant loss of mass. The companion star could be a very young star that has not yet reached its stable configuration and is still

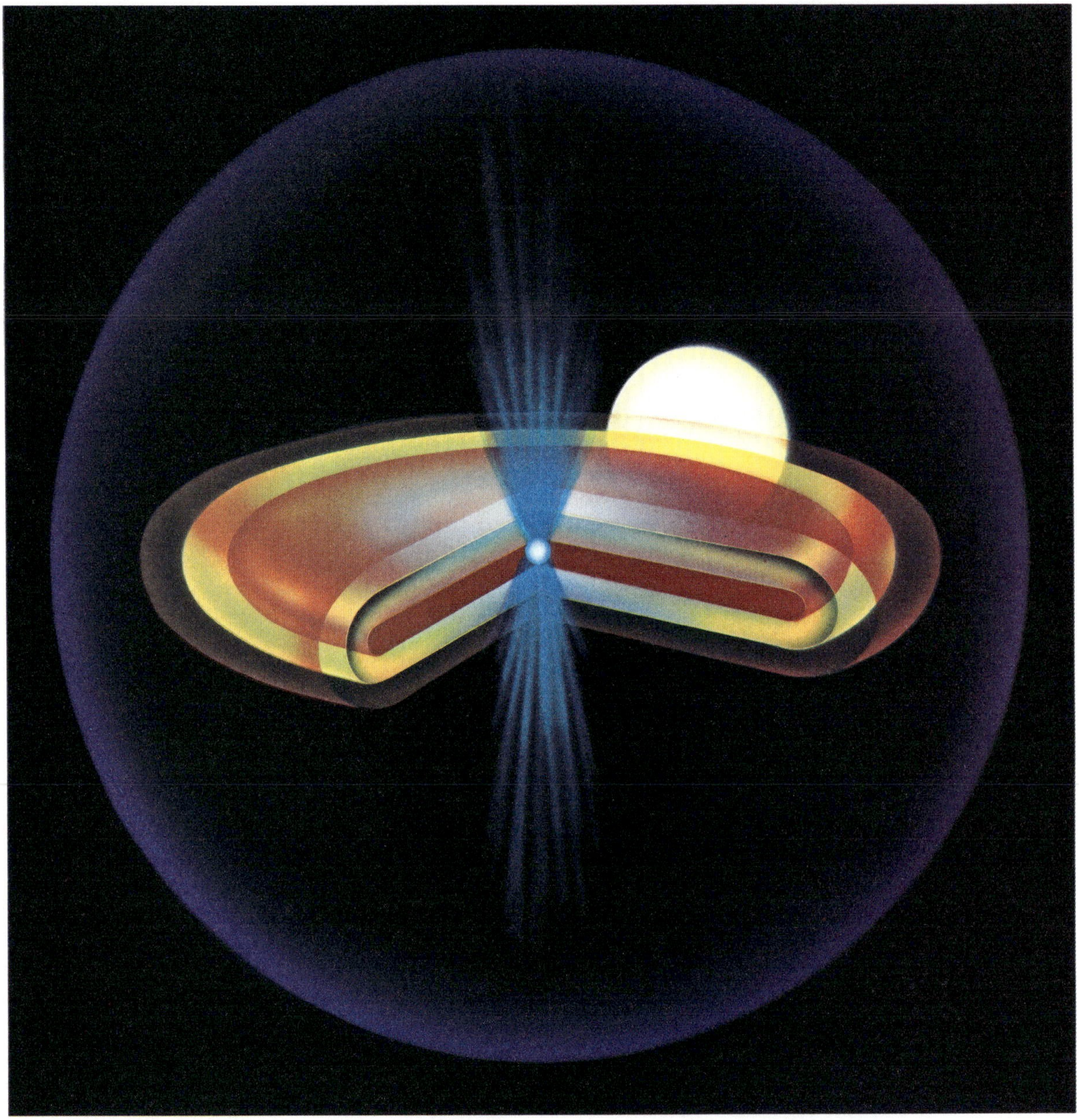

Figure 7.7 MODEL OF EPSILON AURIGAE that tentatively synthesizes the results of infrared, visible and ultraviolet observations. The surface of the supergiant primary star is partly occulted by a ring of dust and several flattened shells of gas that surround a blue-dwarf companion star. A slice has been removed from the eclipsing object to show a cross section. Infrared emission from the ring of dust is shown in reddish brown. Surrounding shells of gas of decreasing density appear as opalescent halos, scattering the yellowish light of the primary and the blue light of the companion star. The parts of the shells of gas most distant from the companion star show the reddish glow of light emitted by the deexcitation of hydrogen. The spherical, outermost envelope of gas is shown as a translucent purple shell.

embedded in the remnant of the dust cloud from which it was formed. The grains of dust in the vicinity of the hot star have a lifetime of less than 10,000 years; if they are not replenished from the outside, the cloud will disappear in this short period of time. The short life expectancy of the cloud and the brief phase of stellar evolution now being traversed by the primary star explain why no other system with the characteristics of Epsilon Aurigae is known: the system offers a glimpse of an exceedingly fleeting stage in the evolution of a binary system.

Verification of these speculations on the nature of the companion star and the evolutionary stage of the system must await a more precise measurement of the masses of the two component stars. The measurement should become possible after the launch of the space telescope. The resolution of its spectrograph will be far greater than that of the instrument carried by the *IUE* satellite. The ultraviolet spectrum of the companion star should exhibit a small Doppler shift, and so a high-resolution spectrum will make it possible to measure its radial velocity as seen from the solar system. With a few hours of observation a year, continuing over a period of 14 years, the radial velocity of the star over half of its orbital period can be determined. From those data the masses of both stars in the system can be derived. Hence by the year 2000 the remaining mystery of Epsilon Aurigae may be completely solved.

A Superluminous Object in the Large Cloud of Magellan

8

A giant nebula in this small galaxy close to our own holds an object that is 50 million times brighter than the sun. If it is one body, it is far more massive than any known star.

• • •

John S. Mathis, Blair D. Savage and Joseph P. Cassinelli
August, 1984

The mass of the sun (2×10^{33} grams) is a standard of measurement for the mass of other celestial objects. Until recently it was the generally accepted view among astronomers that no star much more massive than about 100 times the solar mass could form. Now this view has been challenged by the observations that have been made of a superluminous and possibly supermassive object in a small galaxy close to our own: the Large Cloud of Magellan. The object is designated R136. It is in the Tarantula Nebula, which is also known as the 30 Doradus nebula (see Figure 8.1) because it is in the southern constellation Doradus. If R136 is a single star, it could be as much as 1,000 times more massive than the sun.

Glowing gaseous nebulas are among the loveliest and most impressive objects in the universe. The 30 Doradus nebula is the brightest and largest gaseous nebula in the 30 or so galaxies of the local group that includes our own galaxy. It is of irregular shape and extraordinary size. Whereas to the unaided eye the Great Nebula in Orion looks like a fuzzy star, 30 Doradus covers an area of the sky comparable to that occupied by the sun or the moon, in spite of the fact that it is more than 100 times farther away than the Orion nebula. Its diameter is about 1,000 light-years, compared with the Orion nebula's three. Its gas is highly ionized: most of its atoms have lost one electron or more. Indeed, it contains 1,500 times more ionized gas than the Orion nebula. The ionization must be the result of ultraviolet radiation emitted by the massive hot young stars embedded in the nebula.

R136 (see Figure 8.2) is the brightest object in 30 Doradus. (The designation comes from a catalogue of the brightest stars in the Large Cloud of Magellan prepared by Michael Feast, A. D. Thackeray and A. J. Wesselink of the Radcliffe Observatory in South Africa.) It is near the center of the nebula and is surrounded by dozens of fainter stars, each of which would ordinarily be considered bright. The central object radiates about a million times more light than the sun at visible wavelengths and is another factor of 50 more luminous when ultravio-

Figure 8.1 LARGE CLOUD OF MAGELLAN was photographed with the 61-centimeter Schmidt telescope at the Cerro Tololo Inter-American Observatory in Chile. The Large Cloud is the galaxy closest to our own. The 30 Doradus nebula is the pink area at the center left. The nebula harbors many massive stars hot enough to ionize the interstellar gas; they radiate most of their energy at short wavelengths and appear blue in the photograph. The red patches are ionized gas emitting the red Balmer line of hydrogen in the visible region of the spectrum. The 30 Doradus nebula is the largest such object in the entire local group of galaxies. Some 1,000 light-years in diameter, it occupies an area in the sky that is approximately a third as large as the sun.

Figure 8.2 CLOSE VIEW of the inner region of the 30 Doradus nebula was obtained by John Wood with the four-meter telescope at the Cerro Tololo Observatory. R136 is the tight knot of stars at the center of the photograph. Although it is some 50 million times more luminous than the sun, it is dimmed by interstellar dust. Ultraviolet radiation from R136 ionizes much of the gas in the nebula. R136a and other nearby stars emit powerful stellar winds that have presumably pushed the gas into the arclike surrounding structure visible in the photograph. The central object is either the most massive star known or an extremely dense cluster of massive stars.

let wavelengths are included. Because of its great luminosity, it may account for more ionization than any other object in the nebula.

In 1980 J. V. Feitzinger, Wolfhard Schlosser, Theodor Schmidt-Kaler and Christian Winkler of the University of the Ruhr got excellent photographs of R136 with the 3.6-meter telescope of the European Southern Observatory in Chile. They found that the object actually has at least three distinct components, arranged in a curve something like a comma (see Figure 8.3). The brightest component was designated R136a. R136b and R136c are fainter, redder and probably cooler. The German

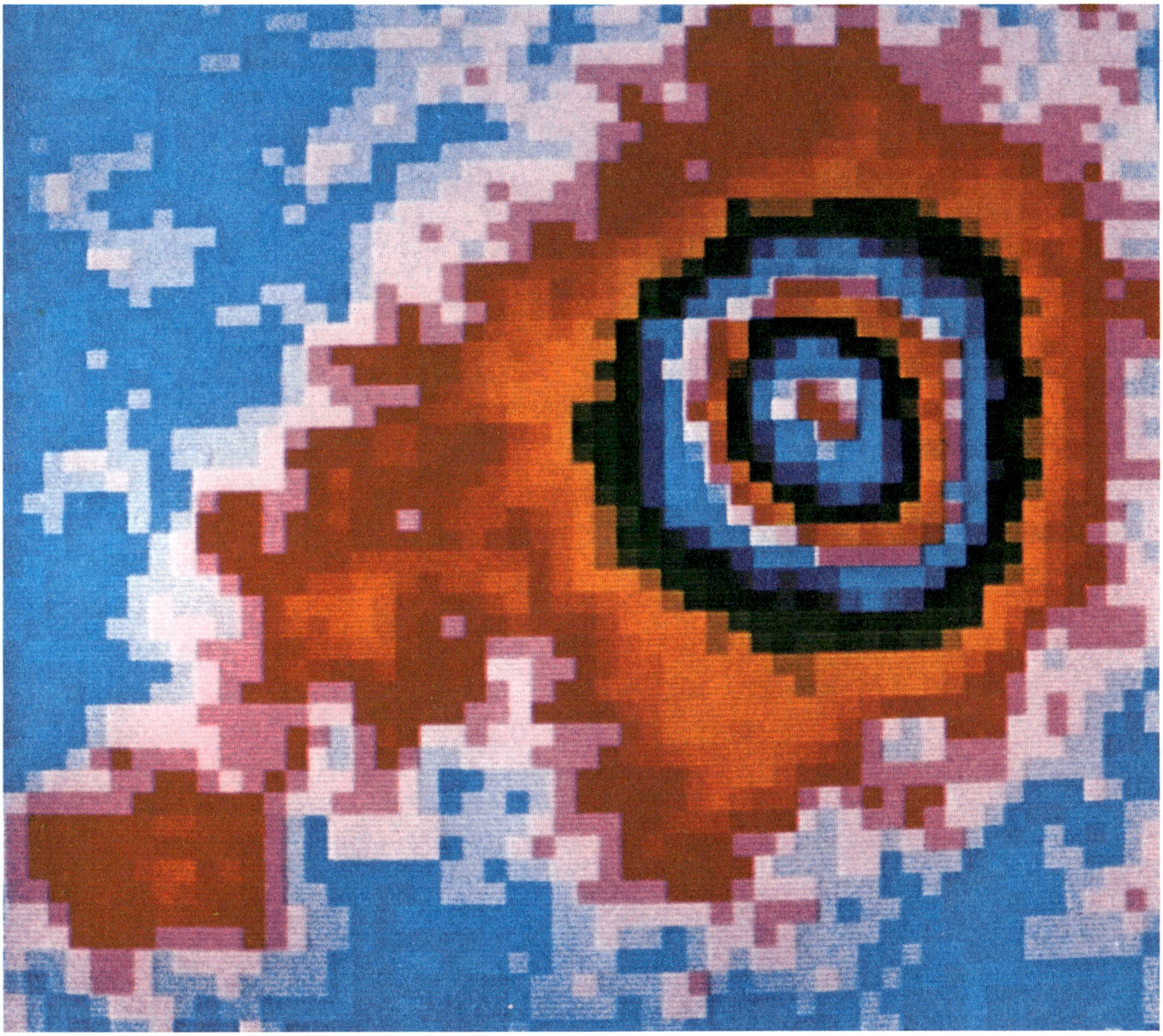

Figure 8.3 AREA OF THE OBJECT R136 appears in this computer-processed photograph made at visible wavelengths. The area is near the center of the Tarantula Nebula (also known as 30 Doradus). The photograph indicates that R136 has at least three components. The brightest one is the irregular, predominantly blue area at the center right; it has been designated R136a. Immediately to its left is R136b, which is about 20 percent as bright as R136a. The reddish area at the lower left is R136c.

workers concluded that R136a might be a star considerably more massive than the commonly accepted upper limit for stellar masses.

The first visible-light spectra of R136, obtained in 1950 by Feast at the Radcliffe Observatory, indicated that the object has a peculiar spectrum characteristic of extremely hot stars. Later work by Nolan R. Walborn of the Space Telescope Science Institute, Peter S. Conti of the University of Colorado, Boulder and Dennis Ebbets of the Space Telescope Institute confirmed that finding.

Stars with a surface temperature higher than 30,000 degrees Kelvin emit most of their radiation at the shorter ultraviolet wavelengths that cannot penetrate the atmosphere of the earth. Therefore ultraviolet observations of R136a were essential to an understanding of the object, but they had to wait for the technical advances brought by the space age. One such advance was the *International Ultraviolet Explorer* satellite (*IUE*), a space observatory operated jointly by the National Aeronautics and Space Administration, the European Space Agency and the Science Research Council of the United Kingdom. With this satellite ultraviolet spectra of R136a were secured in 1978.

The satellite revealed a spectrum resembling in many respects that of the hottest normal stars known, type O3. (The "O" designates the hottest spectral class. The lowest associated numeral designates the hottest star in a class. No O1 or O2 stars have been defined.) It was Walborn who first recognized the O3 stars as a distinct type. They have a surface temperature estimated at 50,00 degrees K. and are among the most luminous stars known.

The ultraviolet spectra of R136a show what are called P Cygni lines, after a highly luminous star in the constellation Cygnus. At ultraviolet wavelengths such lines often originate with highly ionized atoms of carbon, oxygen and nitrogen: C IV, O IV, O V, N IV and N V. (The roman numerals specify the number of electrons removed plus one; for example, C IV is carbon with three electrons removed.) The P Cygni lines have a peculiar profile: they look like an absorption line on the short-wavelength side of the rest wavelength and like an emission line on the long-wavelength side, as can be seen in Figure 8.4. The rest wavelength is one that would be emitted by an object at rest with respect to the observer, and so the P Cygni lines imply that the object is losing mass in the form of an envelope of gas being expelled from it as a stellar wind. The gas on the far side of the object from the observer is going away from him, so that the wavelengths of its emissions are lengthened; the gas on the observer's near side is coming toward him, so that the wavelengths of absorption are shortened (see Figure 8.5).

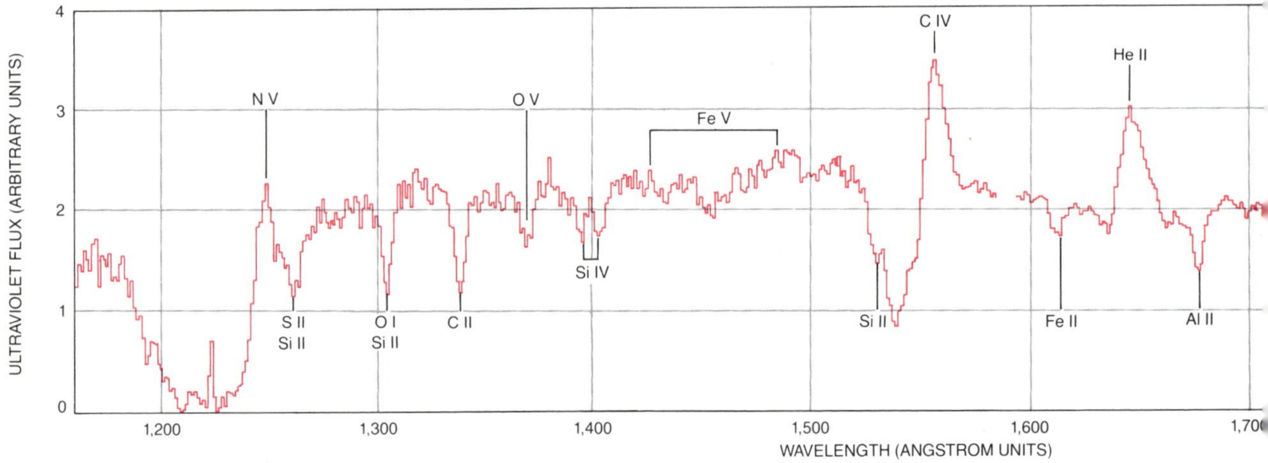

Figure 8.4 ULTRAVIOLET SPECTRUM of R136a plots the intensity of the ultraviolet radiation against wavelength. The emission and absorption features associated with R136a are indicated by the symbols above the spectrum; absorption lines resulting from intervening stellar gas are marked below the spectrum. The symbols represent multiply ionized atoms: nitrogen, oxygen, iron, carbon and helium. The most prominent features are C IV, He II and N IV. The spectra of these ions have a profile characteristic of

The velocity of the outflow of matter from R136a can be estimated from the profile of the P Cygni lines. In particular the wavelengths of the absorption on the short-wavelength side of the C IV line extend 18 angstrom units from the rest wavelength of the line, as shown in Figure 8.6. This indicates that matter is moving away from R136a at a velocity of 3,500 kilometers per second. The stellar wind of R136a is best described as a hurricane; for a normal hot star the average wind speed is about 2,000 kilometers per second. Only type O3 objects exhibit a wind speed as high as that of R136a.

The high surface temperature of R136a is implied by the high state of ionization of the matter in its atmosphere. As the temperature of a star rises its outer layers become more highly ionized because of the increasing number of energetic photons in its radiation and also because collisions between particles become more violent. Thus the presence of some ions and the absence of others as represented in a star's spectrum are diagnostic of temperature. For R136a the presence of lines representing C IV, O V and N V and the absence of lines for Si IV (triply ionized silicon) imply a surface temperature in the range from 45,000 to 80,000 degrees K. In a cooler atmosphere there would be conspicuous lines of Si III and Si IV.

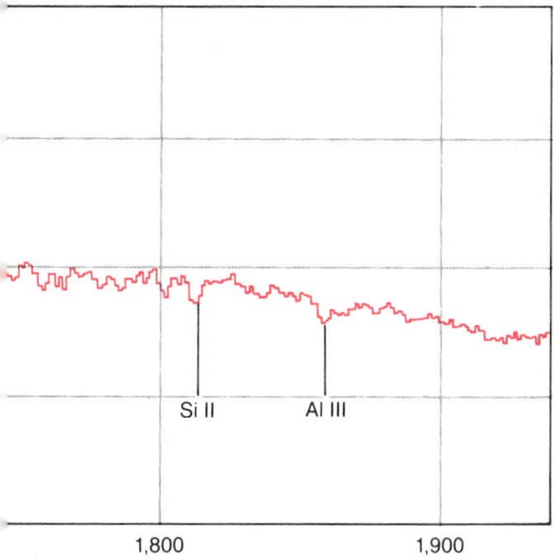

P Cygni: absorption is shifted to shorter wavelengths and emission to longer ones. The high degree of ionization implies a high temperature for the object or objects generating the spectrum.

The ultraviolet spectrum of R136a resembles that of O3 stars except for the exceptionally strong and broad emission line of singly ionized helium (He II) near a wavelength of 1,640 angstroms. Although some O3 stars show He II emission, none exhibit a line as strong and broad as R136a's. It is noteworthy, however, that the peculiar objects known as Wolf-Rayet stars show exceptionally strong He II emission because of their massive stellar wind. These hot stars are thought to be in a stage of evolution more advanced than that of normal O stars. The similarity of the spectrum of R136a to the spectra of both O3 and Wolf-Rayet stars suggests it might even be produced by a collection of stars of both types.

The extreme nature of R136a becomes evident when one considers the number of "normal" stars that would be needed to reproduce its spectrum. For example, R122, whose spectrum is shown in Figure 8.7, is the most luminous normal star in the Large Cloud of Magellan, being three times as bright as the next brightest star in that galaxy. It is three million times more luminous than the sun; most of the other supergiant stars are only 500,000 times more luminous. R122 has an O3 spectrum roughly resembling that of R136 but with weaker He II emission. To get the R136a spectrum from a group of normal stars would call for about 13 stars like R122 or about 40 less luminous O3 stars plus 20 Wolf-Rayet stars of normal brightness. O3 stars are exceedingly rare; only four are known in the entire Large Cloud of Magellan and only 10 in our own galaxy. All these presumably rare objects would have to be packed into a volume with a diameter of one light-year or less.

In terms of stellar evolution it is not too unreasonable to suppose a mixture of O3 and Wolf-Rayet stars could be identified; indeed, such a mixture exists in the giant nebula in our own galaxy called the Carina Nebula. The four O3 stars in the Carina Nebula, however, extend in space over a distance of 10 light-years.

One can try to determine the nature of R136a by studying the structure of the R136 region on photographs made with telescopes on the ground. The main trouble with this approach is that the earth's atmosphere blurs the image. Nevertheless, at a time of excellent "seeing" last year Y. H. Chu of the University of Wisconsin at Madison made photographs of the inner region of 30 Doradus in order

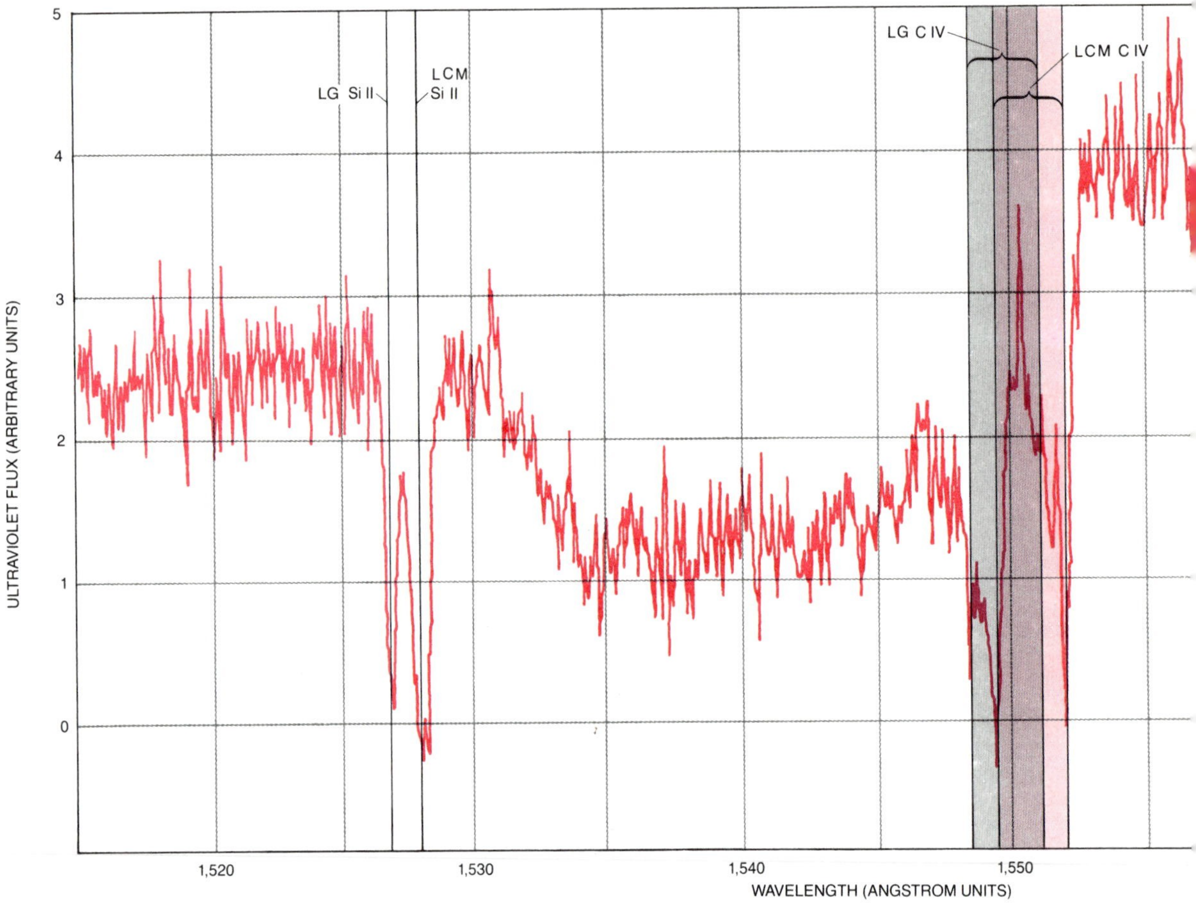

Figure 8.5 HIGHER RESOLUTION of the spectrum of R136a was obtained with the high-resolution spectrograph of the *International Ultraviolet Explorer*. The focus is on a line in the spectrum of C IV. Various narrow absorption lines that originate in the interstellar gas of our own galaxy (*LG*, for local galaxy) and of the Large Cloud of Magellan (*LCM*) are marked. The broad C IV feature at the right, originating in R136, is an absorption line at shorter wavelengths and an emission line at longer ones. The absorption at shorter wavelengths implies, on the basis of the

to sort out the types and numbers of stars near R136.

Many hot stars lie within several minutes of arc of R136, and Jorge Melnick of the University of Chile had identified several of them as stars of type O3. Chu has carefully analyzed many images of the R136 region and has concluded that within the R136a component, which has a diameter of only three seconds of arc, one can identify at least four stellar objects. Unfortunately the instruments of the *International Ultraviolet Explorer* can resolve individual stars only if they are separated by at least three seconds of arc, and so the spectra from the satellite include contributions from all sources of ultraviolet radiation in the R136a area. Chu has labeled the dominant source R136a1. It is this object that is now the candidate for a superluminous star. Chu has also identified a fainter point source, R136a2, about .5 arc-second from a1.

A technique that can be exploited to determine the details of an image distorted by the earth's atmosphere is speckle interferometry. The technique

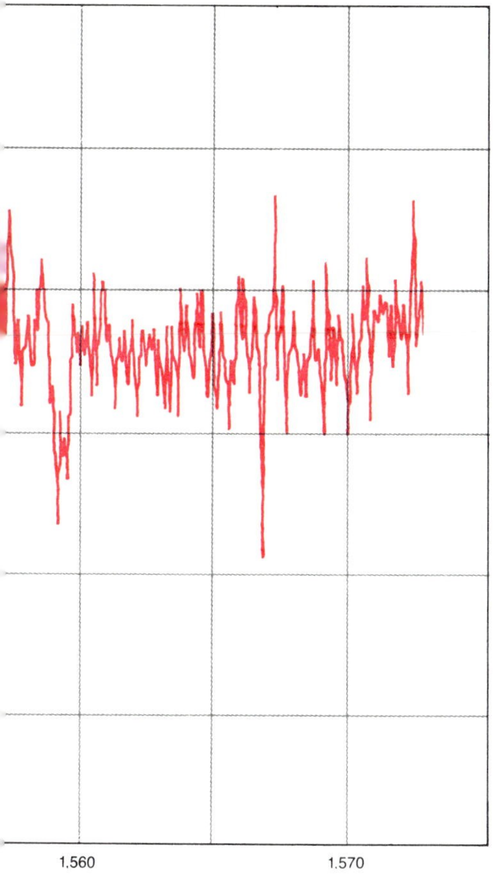

Doppler effect, that matter is flowing outward from R136 at the high velocity of 3,600 kilometers per second. The spectrum is a profile of the characteristic P Cygni type.

consists in making hundreds of very short exposures in rapid succession. Each exposure shows the image when the atmosphere is in a definite (but unknown) state. By applying certain mathematical procedures to the individual images the investigator can extract the true structure of the image almost as well as the mirror of the telescope could resolve it if the telescope were above the atmosphere.

Gerd Weigelt of the University of Erlangen-Nuremberg applied this technique to images of R136a1 made with the European Southern Observatory's 3.6-meter telescope. At about the same time John Meaburn of the University of Manchester and his colleagues got somewhat different results with the 3.9-meter Anglo-Australian telescope in Australia. Presumably the difference arises from the way the two groups analyzed the images.

The speckle-interferometry results agree on the fact that R136a1 consists of a dominant unresolved object with several fainter objects close to it. Meaburn believes nothing else in the area is of comparable brightness, but Weigelt maintains that a second unresolved object is about .5 arc-second away (in the same position where Chu had placed R136a2) and is about a fifth as bright as R136a1 at visible wavelengths. In addition Charles E. Worley of the United States Naval Observatory has confirmed the position of R136a2 by visual observation. We therefore believe the reality of R136a2 has probably been established.

Weigelt's speckle-interferometry results suggest that the morphology of R136 is even more complex. He finds another component, which Chu calls a3, only .1 arc-second from a1—too close for Chu to resolve it. The speckle-interferometry results do not help much in determining relative brightness, but a3 seems to be comparable to a2, that is, it is about a fifth as bright as a1. All observers who have been studying R136 from the ground agree that there is also a background of still fainter stars within its three-arc-second image.

With Weigelt's results Chu was able to estimate the relative brightness of the three components. The light from the dominant component, a1, is equivalent to that from six stars like R122 or perhaps 20 of the more typical O3 stars. According to Weigelt, the angular diameter of a1 is no more than .08 arc-second, or about 24 light-days. By astronomical standards that is a small region. For example, the star nearest the sun is four light-years away. Nevertheless, the radius of the 24-light-day volume is 55 times the distance from the sun to Pluto. In that space there is certainly room for six stars or even 20.

Is R136a1 a single star six times more luminous than any now known or is it a cluster of stars that are individually like, say, R122? The question is of great interest because either hypothesis extends astronomy to objects of larger mass and greater luminosity than anything now recognized.

If R136a1 is a single star, it must have a mass of between 400 and 1,000 solar masses, making it at least twice as massive as any previously known star.

The estimate of mass comes rather directly from luminosity. The outer layers of the atmosphere of a star must be held down by gravity against the pressure of the emerging radiation. The minimum mass for a star of a given luminosity was established by the British astronomer A. S. Eddington some 60 years ago. A star cannot violate the Eddington limit and be in even approximate mechanical equilibrium.

In 1970 Franz D. Kahn of the University of Manchester showed that very massive stars may not be able to form out of the clouds of cold gas and dust in which all stars originate. The problem is that as the cloud collapses, the density in the center increases much faster than the density in the outer regions. The core heats up, its dust grains are sublimated and its gas is ionized. It becomes a luminous, continuously growing object, radiating away approximately half of the gravitational energy converted into heat in its collapse. The result is a protostar surrounded by a dust-free zone surrounded in turn by a dusty shell.

In the dusty shell the light and the ultraviolet radiation from the protostar are converted into infrared radiation characteristic of the lower temperature at which dust can exist (about 1,000 degrees K.). The infrared radiation is absorbed in the outermost parts of the collapsing cloud, where the infalling gas and dust are only loosely bound by gravity. The outward-directed momentum of the absorbed radiation tends to reverse the infall of the dust grains. The now outflowing dust in turn drags along its gas. The growth of the star stops.

Hence there is an upper limit (only 40 solar masses in Kahn's original paper) on the central-star mass that will allow mass to continue falling inward. The Kahn model is sensitive, however, to the properties assumed for the dust, as one might suspect from the central role dust plays in the reversal of flow. Our University of Wisconsin colleague Mark G. Wolfire has shown that if Kahn had assumed a dust-sublimation temperature of 2,000 degrees K. instead of 3,600, the upper limit on the mass of the protostar could have been raised to about 1,000 solar masses, keeping all the other approximations of the model unchanged. The lower sublimation temperature is reasonable; Edward P. Ney and his colleagues at the University of Minnesota have found that in the expanding shell of gas around a nova dust condenses when the gas cools to a temperature of about 2,000 degrees. There are also other processes for the destruction of grains, such as collisions with helium atoms as the dust is pushed through the gas by radiation, that could make the collapse of a larger mass possible.

The original dust content per gram of the protostellar cloud is also important. In our own galaxy the ratio of dust to gas in interstellar matter tends to be large in the central plane and in regions of the galaxy closer to the center than the sun is. Toward the outer regions of the galaxy, however, the abundance of dust decreases.

A supermassive star would be unlikely to form in a region that has a large amount of dust, since radiation pressure pushing against the dust would tend to prevent the necessary collapse of a large dust cloud. The ratio of dust to gas near 30 Doradus has been found to be about a third of the ratio in the neighborhood of the sun. It is therefore not unreasonable to suppose a star of 400 or more solar masses could have formed in the 30 Doradus region.

The structure of a star is determined by several balancing processes, such as an equilibrium between the inward force of gravity and the outward force provided by pressure. In a very massive star the rate at which nuclear energy is generated is quite sensitive to the temperature at the center. Slight perturbations have a tendency to grow. In addition the pressure at the center is primarily the pressure not of gas but of radiation. A star supported entirely by radiation can easily be dispersed; in fact, with radiation and gravity in balance it can disperse without any further input of energy. It is only the small fraction of the pressure arising from

Figure 8.6 P CYGNI SPECTRAL PROFILE is formed in the expanding atmosphere (the stellar wind) surrounding a star that is losing mass (*a*). The shortward-displaced absorption (the decreased flux of radiation at wavelengths shorter than λ_0, the wavelength of the line if the atmosphere were not expanding, as shown in *b*) is caused by photons scattered out of the line of sight by ions in the blue region on the near side of the star. In a spherically symmetrical outflow every photon scattered out of the line of sight is matched by one scattered into it by ions in the red emission lobe. Photons scattered toward the observer from behind the star (*black region*) are not seen because they are blocked by the star. The P Cygni profile (*c*) results from a combination of three effects: scattering out of the line of sight from the blue region, scattering into the line of sight from the red region and the absence of photons from behind the star.

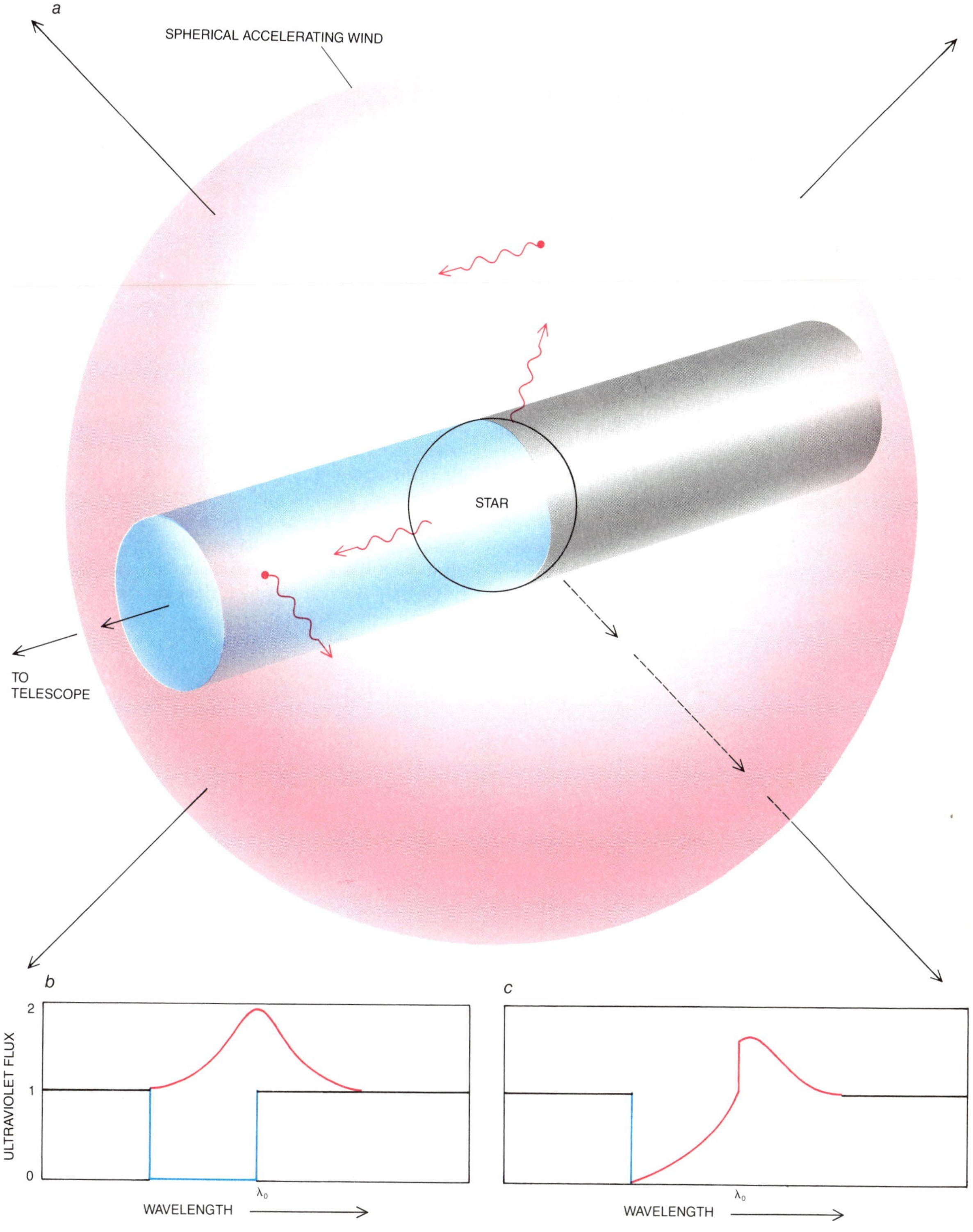
a
SPHERICAL ACCELERATING WIND
STAR
TO
TELESCOPE
b
c
ULTRAVIOLET FLUX
2
1
0
λ0
WAVELENGTH
λ0
WAVELENGTH

the thermal motion of the gas that stabilizes the massive star. Calculations have suggested that stars more massive than about 60 suns might be dispersed by internally driven pulsations.

Exceptionally massive stars can be described mathematically in a fairly simple way. In 1962 Fred Hoyle of the University of Cambridge and William A. Fowler of the California Institute of Technology derived the relevant equations. (They were interested in objects that might have a mass of up to a million solar masses, which they had proposed to explain the recently discovered quasistellar objects, or quasars.) The interior of a massive star is well mixed by rising and sinking convection currents. In a star smaller than 60 solar masses the interior mixing can lead to a change in chemical composition from the interior layers, where energy is transported by convection, to the outer layers, where energy is transported by radiation.

André Maeder of the Geneva Observatory has shown that stars larger than 60 solar masses should be almost homogeneous because their outer layers are removed rapidly by stellar winds. In such a star evolution proceeds straightforwardly. At the beginning the star's nuclear fuel is hydrogen, which in the course of thermonuclear burning is converted into helium. The composition of the star therefore gradually changes from being about 70 percent hydrogen to being primarily helium. In the process a homogeneous star decreases in radius and increases in temperature. Stars of lower mass, in contrast, evolve from relatively compact hydrogen stars to cooler and larger giant stars. The surface temperature of a massive star increases from about 60,000 degrees K. to about 90,000 degrees as it ages and becomes a helium star.

The absence of the usually strong Si IV lines in the ultraviolet spectrum of R136a1 is evidence for a gas temperature consistent with the hypothesis that the object is a single star. One might expect, however, that a very massive star would be variable in its output of light (and thus unstable) because of the marginal stability of its internal structure. Richard Stothers of the Institute for Space Studies of NASA has investigated several processes that can in principle bring about stability in such a star. They include high-speed rotation and the entanglement of lines of force in interior magnetic fields. Assuming processes of this kind to be at work in R136a1, the ultraviolet spectroscopic data are consistent with the single-star hypothesis. The data do not, however, rule out other possibilities.

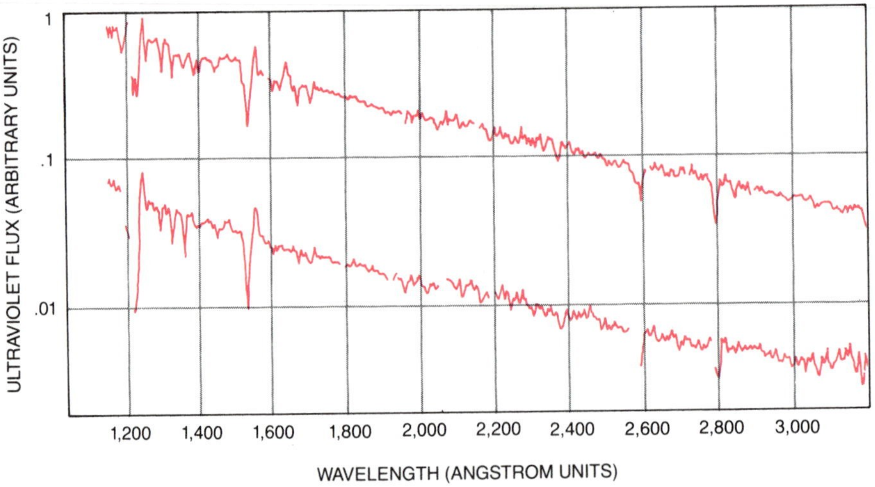

Figure 8.7 ULTRAVIOLET FLUX of R136a is charted (*upper curve*) as it is measured from above the earth's atmosphere. The range of wavelengths was observed by the *International Ultraviolet Explorer* spectrometer, and the reading was corrected for attenuation by dust in our galaxy and the Large Cloud of Magellan. For comparison the flux distribution of R122, a type O3 star at the same distance as R136a, is also shown (*lower curve*). The spectral lines are similar except for the helium-ion line (He II), at 1,640 angstroms. R136a is, however, about 13 times brighter than R122, which is the most luminous "normal" star in the Large Cloud of Magellan.

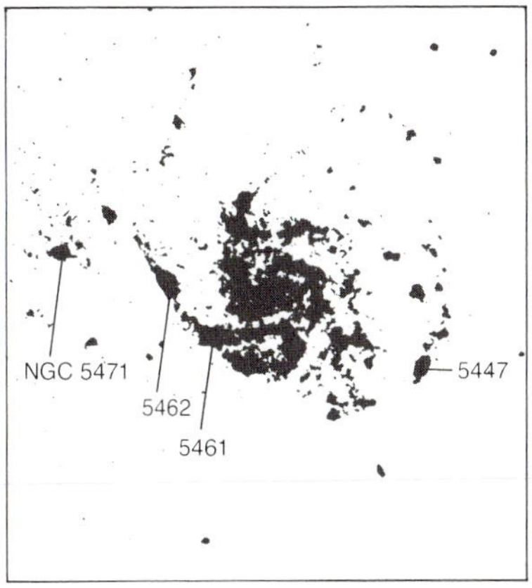

Figure 8.8 ULTRAVIOLET IMAGE of a distant galaxy, the supergiant system M101, was made from a high-altitude rocket in a project carried out by T. P. Stecher and R. C. Bohlin of the Goddard Space Flight Center of the National Aeronautics and Space Administration. The image emphasizes the hot stars that ionize the gas around them. In the accompanying map the four lines point to nebulas that are all at least five times more luminous than 30 Doradus. The nature of the sources that excite the radiation of these nebulas is not clear. Moreover, it is difficult to determine because of their distance (about 20 million light-years). The spectra obtained by the *International Ultraviolet Explorer* nonetheless suggest that the four sources resemble R136a.

Anthony F. J. Moffat of the University of Montreal and Wilhelm Seggewiss of the Hoher List Observatory in West Germany believe R136a1 is a tight cluster of stars, each of a mass comparable to the mass of stars already known. Probably the most massive individual stars known today are R122 and Eta Carina, an unstable object surrounded by a thick cloud of dust. Stellar theory predicts that each of these stars has about 200 times the mass of the sun. As we have pointed out, a cluster supplying the energy of R136a1 would require at least six R122's within a remarkably small volume.

That hypothesis raises two questions. Would not the first star, as soon as it had formed, prevent the formation of others by heating the collapsing gas? How can so much gas get into such a small volume without having collapsed earlier? To date these questions have no answers, but the same thing can be said of many other puzzles associated with the formation of stars.

Chu's visible-light images indicate that the bright component, R136a1, is near the center of a compact cluster of stars. This observation raises the interesting possibility that R136a1 has formed as the consequence of the dynamical evolution of a star cluster. In regions with a high density of stars gravitational encounters between stars can result in the evaporation, or loss, of low-mass stars, causing the cluster to shrink. Eventually supermassive stars might be created by the actual physical coalescence of stars in such encounters. The theory of the dynamical evolution of clusters has received considerable attention in recent years because of its possible application to the theory of the origin of the peculiar luminous objects found in the nucleus of galaxies. It is exciting to think R136a1 might represent an example of stellar coalescence in a galaxy next door to our own.

So far we have discussed R136a1 with the implied assumption that the object burns hydrogen like an ordinary star. Perhaps we should consider certain more exotic possibilities. For example, if black holes exist, the interstellar matter spiraling into one of them would form an accretion disk. The process would generate large quantities of radiation, giving rise to the kind of brightness observed in R136a1. Another exotic possibility is the spinar, a rotating mass of ionized and magnetized gas that has been put forward as an explanation of the large quantities of radiation emitted by quasars.

Our strongest reason for not proposing such hypotheses for R136a1 is that the observational data now available can be explained by a fairly straightforward extrapolation from massive hydrogen-burning stars on the main sequence of stellar evolution. The distribution of energy output across the spectrum of R136a1 does not perfectly match current theoretical models of other stars, but the differences are no worse than those encountered with the luminous type-O stars and are probably due to the effects of massive stellar winds. The velocity of the R136a1 wind is high, but it is not out of line with that of the wind from ordinary O3 stars. The rate at which R136a1 loses mass is huge (one earth mass per week), but even that is in line with an extrapolation from type-O and Wolf-Rayet stars. Finally, one would expect black holes and other exotic objects to give rise to a large flux of X rays. Knox S. Long, Jr., of Johns Hopkins University, working with the Einstein observatory satellite, has observed X rays from objects in the center of 30 Doradus. Although the X-ray output from R136a and its environment is large (equivalent to about 100 times the total energy output of the sun), it too is not significantly out of line with an extrapolation from O stars.

Whatever the true nature of R136a is, it is almost certainly not a unique object. In our galaxy there is a giant ionized region (NGC 3603) with a luminosity about a seventh that of 30 Doradus. Walborn first noted the strong similarity between the core object of NGC 3603 and R136a. Perhaps the former is a scaled-down version of the latter.

One must go to galaxies beyond the Large Cloud of Magellan to find objects that rival the 30 Doradus nebula. Frank Israel of the Leiden Observatory and others have studied the largest nebulas in galaxies out to about 25 million light-years. Within that volume are a few spiral galaxies with notably active regions where hot, massive stars are forming and the associated ionized nebulas can be perceived. Israel lists seven other galaxies, all at least 10 million light-years away, with nebulas more luminous than 30 Doradus. The most spectacular is the galaxy M101, which harbors four nebulas that are more than five times as luminous as 30 Doradus (see Figure 8.8). One of them is NGC 5461, the most luminous nebula known; it is the equivalent of 11 nebulas like 30 Doradus. Little is known about the sources that excite these huge gas complexes; they are about 10 times farther away than 30 Doradus, meaning that the earth receives only about a hundredth as much radiation from their individual stars as it does from the stars of 30 Doradus.

Philip L. Massey and John B. Hutchings of the Dominion Astrophysical Observatory in British Columbia have worked with the *International Ultraviolet Explorer* satellite to examine the luminous stars in the large ionized regions of the spiral galaxy M33 in the local group of galaxies. Six of the seven objects they studied have ultraviolet spectra quite similar to R136a's. Unfortunately M33 is 10 times farther away than R136, so that it is impossible to tell whether the spectra represent superluminous stars or highly compact groups of stars with normal characteristics.

The Space Telescope that is to be launched by NASA should help to clarify the uncertainties about the physical nature of R136a. With an angular resolution 10 times better than can be achieved from the ground and with ultraviolet-spectroscopic capabilities, the Space Telescope will determine the relative ultraviolet brightness of the various components of R136a. The high stability of the telescope for purposes of measurement will make possible a careful search for variability in R136a. The detection of such variability would limit the range of possible explanations for this bizarre object. The telescope will also be employed to study the central objects of other supergiant nebulas beyond 30 Doradus. It may turn out that objects as unusual as R136a lie at the core of most giant ionized regions in other galaxies.

SECTION

IV

SUPERNOVAS AND REMNANTS

. . .

How a Supernova Explodes

9

When a large star runs out of nuclear fuel, the core collapses in milliseconds. The subsequent "bounce" of the core generates a shock wave so intense that it blows off most of the star's mass.

• • •

Hans A. Bethe and Gerald Brown

May, 1985

The death of a large star is a sudden and violent event. The star evolves peacefully for millions of years, passing through various stages of development, but when it runs out of nuclear fuel, it collapses under its own weight in less than a second. The most important events in the collapse are over in milliseconds. What follows is a supernova, a prodigious explosion more powerful than any since the big bang with which the universe began.

A single exploding star can shine brighter than an entire galaxy of several billion stars. In the course of a few months it can give off as much light as the sun emits in a billion years. Furthermore, light and other forms of electromagnetic radiation represent only a small fraction of the total energy of the supernova. The kinetic energy of the exploding matter is 10 times greater. Still more energy—perhaps 100 times more than the electromagnetic emission—is carried away by the massless particles called neutrinos, most of which are emitted in a flash that lasts for about a second. When the explosion is over, most of the star's mass has been scattered into space, and all that remains at the center is a dense, dark cinder. In some cases even that may disappear into a black hole.

Such an outline description of a supernova could have been given almost 30 years ago, and yet the detailed sequence of events within the dying star is still not known with any certainty. The basic question is this: A supernova begins as a collapse, or implosion; how does it come about, then, that a major part of the star's mass is expelled? At some point the inward movement of stellar material must be stopped and then reversed; an *im*plosion must be transformed into an *ex*plosion.

Through a combination of computer simulation and theoretical analysis a coherent view of the supernova mechanism is beginning to emerge. It appears the crucial event in the turnaround is the formation of a shock wave that travels outward at 30,000 kilometers per second or more.

Supernovas are rare events. In our own galaxy just three have been recorded in the past 1,000 years; the brightest of these, noted by Chinese observers in 1054, gave rise to the expanding shell of gas now known as the Crab Nebula. If only such nearby events could be observed, little would be known about supernovas. Because they are so luminous, however, they can be detected even in distant galaxies, and 10 or more per year are now sighted by astronomers.

The first systematic observations of distant supernovas were made in the 1930's by Fritz Zwicky of the California Institute of Technology. About half of the supernovas Zwicky studied fitted a quite consistent pattern: the luminosity increased steadily for about three weeks and declined gradually over a period of six months or more. He designated the explosions in this group Type I. The remaining supernovas were more varied, and Zwicky divided them into four groups; today, however, they are all grouped together as Type II. In Type I and Type II supernovas the events leading up to the explosion are thought to be quite different. Here we shall be concerned primarily with Type II supernovas.

The basis for the theory of supernova explosions was the work of Fred Hoyle of the University of Cambridge. The theory was then developed in a fundamental paper published in 1957 by E. Margaret Burbidge, Geoffrey R. Burbidge and William A. Fowler, all of Caltech, and Hoyle. They proposed that when a massive star reaches the end of its life, the stellar core collapses under the force of its own gravitation. The energy set free by the collapse expels most of the star's mass, distributing the chemical elements formed in the course of its evolution throughout interstellar space. The collapsed core leaves behind a dense remnant, in many cases a neutron star.

A supernova is an unusual and spectacular outcome of the sequence of nuclear fusion reactions that is the life history of a star (see Figure 9.1). The heat given off by the fusion creates pressure, which counteracts the gravitational attraction that would otherwise make the star collapse. The first series of fusion reactions have the net effect of welding four atoms of hydrogen into a single atom of helium. The process is energetically favorable: the mass of the helium atom is slightly less than the combined masses of the four hydrogen atoms, and the energy equivalent of the excess mass is released as heat.

The process continues in the core of the star until the hydrogen there is used up. The core then contracts, since gravitation is no longer opposed by energy production, and as a result both the core and the surrounding material are heated. Hydrogen fusion then begins in the surrounding layers. Meanwhile the core becomes hot enough to ignite other fusion reactions, burning helium to form carbon, then burning the carbon to form neon, oxygen and finally silicon. Again each of these reactions leads to the release of energy. One last cycle of fusion combines silicon nuclei to form iron, specifically the common iron isotope ^{56}Fe, made up of 26 protons and 30 neutrons. Iron is the end of the line for spontaneous fusion. The ^{56}Fe nucleus is the most strongly bound of all nuclei, and further fusion would absorb energy rather than releasing it.

At this stage in the star's existence it has an onionlike structure (see Figure 9.2). A core of iron and related elements is surrounded by a shell of silicon and sulfur, and beyond this are shells of oxygen, carbon and helium. The outer envelope is mostly hydrogen.

Only the largest stars proceed all the way to the final, iron-core stage of the evolutionary sequence. A star the size of the sun gets no further than helium burning, and the smallest stars stop with hydrogen fusion. A larger star also consumes its stock of fuel much sooner, even though there is more of it to begin with; because the internal pressure and temperature are higher in a large star, the fuel burns faster. Whereas the sun should have a lifetime of 10 billion years, a star 10 times as massive can complete its evolution 1,000 times faster. Regardless of how long it takes, all the usable fuel in the core will eventually be exhausted. At that point heat production in the core ends and the star must contract.

When fusion ends in a small star, the star slowly shrinks, becoming a white dwarf: a burned-out star that emits only a faint glow of radiation. In isolation the white dwarf can remain in this state indefinitely, cooling gradually but otherwise changing little. What stops the star from contracting further? The answer was given more than 50 years ago by Subrahmanyan Chandrasekhar of the University of Chicago.

Loosely speaking when ordinary matter is compressed, higher density is achieved by squeezing out the empty space between atoms. In the core of a white dwarf this process has reached its limit: the atomic electrons are pressed tightly together. Under these conditions the electrons offer powerful resistance to further compression.

Chandrasekhar showed there is a limit to how much pressure can be resisted by the electrons' mutual repulsion. As the star contracts, the gravitational energy increases, but so does the energy of the electrons, raising their pressure. If the contraction goes very far, both the gravitational energy and the electron energy are inversely proportional to the

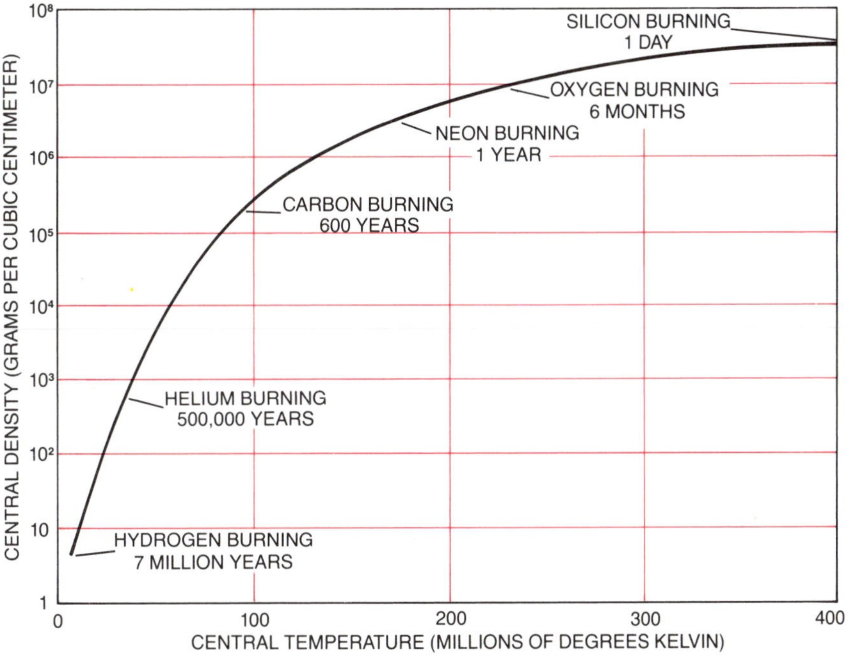

Figure 9.1 EVOLUTION OF A MASSIVE STAR is a steadily accelerating progress toward higher temperature and density in the core. When the hydrogen in the core is exhausted, the core contracts, which heats it enough to ignite the fusion of helium into carbon. This cycle then repeats, at a steadily increasing pace, through the burning of carbon, neon, oxygen and silicon. The final stage of silicon fusion yields a core of iron, from which no further energy can be extracted by nuclear reactions. Hence the iron core cannot resist gravitational collapse, leading to a supernova explosion. (Data in this figure and Figure 9.2 are based on calculations by Tom Weaver.)

star's radius. Whether or not there is some radius at which the two opposing forces are in balance, however, depends on the mass of the star. Equilibrium is possible only if the mass is less than a critical value, now called the Chandrasekhar mass. If the mass is greater than the Chandrasekhar limit, the star must collapse.

The value of the Chandrasekhar mass depends on the relative numbers of electrons and nucleons (protons and neutrons considered collectively): the higher the proportion of electrons, the larger the electron pressure and so the larger the Chandrasekhar mass. In small stars where the chain of fusion reactions stops at carbon the ratio is approximately ½ and the Chandrasekhar mass is 1.44 solar masses. This is the maximum stable mass for a white dwarf.

A white dwarf with a mass under the Chandrasekhar limit can remain stable indefinitely; nevertheless, it is just such stars that are thought to give rise to Type I supernovas. How can this be? The key to the explanation is that white dwarfs that explode in supernovas are not solitary stars but rather are members of binary star systems. According to one hypothesis, matter from the binary companion is attracted by the intense gravitational field of the dwarf star and gradually falls onto its surface, increasing the mass of the carbon-and-oxygen core. Eventually the carbon ignites at the center and burns in a wave that travels outward, destroying the star.

The idea that explosive carbon burning triggers Type I supernovas was proposed in 1960 by Hoyle and Fowler. More detailed models have since been devised by many astrophysicists, most notably Icko Iben, Jr., and his colleagues at the University of Illinois at Urbana-Champaign. Recent calculations done by Ken'ichi Nomoto and his colleagues at the University of Tokyo suggest that the burning is actually not explosive. The wave of fusion reactions propagates like the burning of a fuse rather than

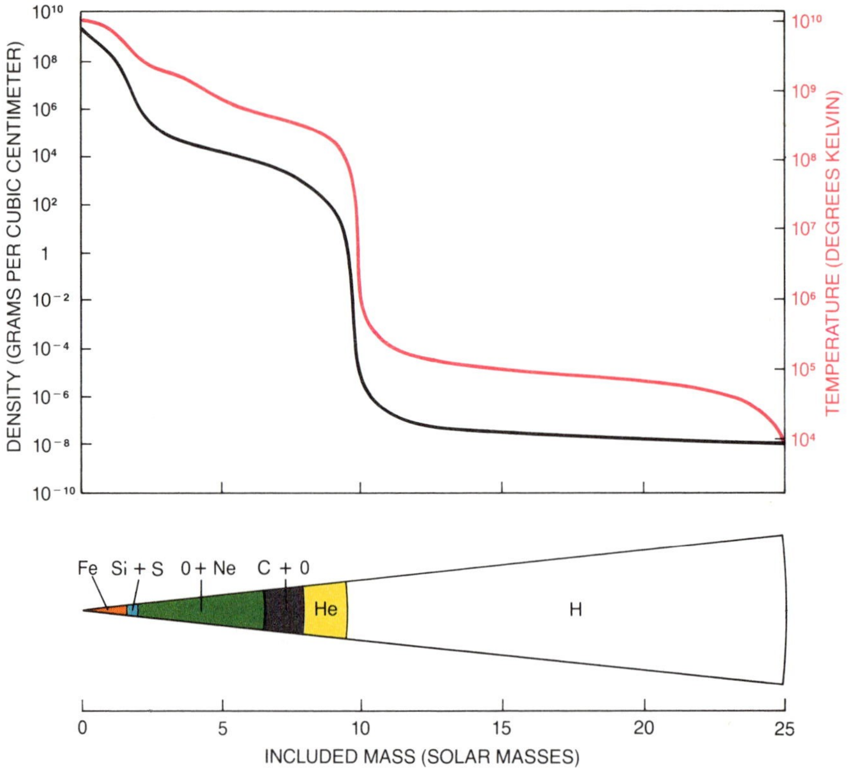

Figure 9.2 ONIONLIKE STRUCTURE is characteristic of a massive star at the end of its evolution, just before the gravitational collapse. The iron core is embedded in a mantle of silicon, sulfur, oxygen, neon, carbon and helium, surrounded by an attenuated envelope of hydrogen. Temperature and density fall off steadily in the mantle, then drop precipitously at the envelope. Fusion has stopped in the core but continues at boundaries between layers.

like the explosion of a firecracker; it is a deflagration rather than a detonation.

Even though the burning is less violent than a detonation, the white dwarf is completely disrupted. The initial binding energy that holds the star together is approximately 10^{50} ergs; the energy released by the burning is 20 times greater (2×10^{51} ergs), enough to account for the 10,000-kilometer-per-second velocity of supernova remnants. In the course of the deflagration nuclear reactions create about one solar mass of the unstable nickel isotope ^{56}Ni, which decays into ^{56}Co and then ^{56}Fe over a period of months. The rate of energy release from the radioactive decay is just right to account for the gradually declining light emission from Type I supernovas.

The Type II supernovas that are our main concern here arise from much more massive stars. The lower limit is now thought to be about eight solar masses.

In tracing the history of a Type II supernova it is best to begin at the moment when the fusion of silicon nuclei to form iron first becomes possible at the center of the star. At this point the star has already passed through stages of burning hydrogen, helium, neon, carbon and oxygen, and it has the onionlike structure described above. The star has taken several million years to reach this state. Subsequent events are much faster.

When the final fusion reaction begins, a core made up of iron and a few related elements begins to form at the center of the star, within a shell of silicon. Fusion continues at the boundary between the iron core and the silicon shell, steadily adding mass to the core. Within the core, however, there is no longer any production of energy by nuclear reactions; the core is an inert sphere under great pressure. It is thus in the same predicament as a white dwarf: it can resist contraction only by electron pressure, which is subject to the Chandrasekhar limit.

Once the fusion of silicon nuclei begins, it proceeds at an extremely high rate, and the mass of the core reaches the Chandrasekhar limit in about a day. We noted above that for a white dwarf the Chandrasekhar mass is equal to 1.44 solar masses; for the iron core of a large star the value may be somewhat different, but it is probably in the range between 1.2 and 1.5 solar masses (see Figure 9.3).

When the Chandrasekhar mass has been attained, the pace speeds up still more. The core that was built in a day collapses in less than a second. The task of analysis also becomes harder at this point, so that theory relies on the assistance of computer simulation. Computer programs that trace the evolution of a star have been developed by a number of workers, including W. David Arnett of the University of Chicago and a group at the Lawrence Livermore National Laboratory led by Tom Weaver of that laboratory and Stan Woosley of the University of California, Santa Cruz. They are the "burners" of stars; we and our colleagues in theoretical physics are "users" of their calculations.

The simulations furnish us with a profile of the presupernova core, giving composition, density and temperature as a function of radius. The subsequent analysis relies on applying familiar laws of thermodynamics, the same laws that describe such ordinary terrestrial phenomena as the working of a heat engine or the circulation of the atmosphere.

It is worthwhile tracing in some detail the initial stages in the implosion of the core. One of the first points of note is that compression raises the temperature of the core, which might be expected to raise the pressure and slow the collapse. Actually the heating has just the opposite effect.

Pressure is determined by two factors: the number of particles in a system and their average energy. In the core both nuclei and electrons contribute to the pressure, but the electron component is much larger. When the core is heated, a small fraction of the iron nuclei are broken up into smaller nuclei, increasing the number of nuclear particles and raising the nuclear component of the pressure. At the same time, however, the dissociation of the nuclei absorbs energy; since energy is released when an iron nucleus is formed, the same quantity of energy must be supplied in order to break the nucleus apart. The energy comes from the electrons and decreases their pressure. The loss in electron pressure is more important than the gain in nuclear pressure. The net result is that the collapse accelerates.

It might seem that the implosion of star would be a chaotic process, but in fact it is quite orderly. Indeed, the entire evolution of the star is toward a condition of greater order, or lower entropy. It is easy to see why. In a hydrogen star each nucleon can move willy-nilly along its own trajectory, but in an iron core groups of 56 nucleons are bound together and must move in lockstep. Initially the entropy per nucleon, expressed in units of Boltzmann's constant, is about 15; in the presupernova core it is less than 1. The difference in entropy has been carried off during the evolution of the star by electromagnetic radiation and toward the end also by neutrinos.

The low entropy of the core is maintained throughout the collapse. Nuclear reactions continually change the species of nuclei present, which one might think could lead to an increase in entropy; the reactions are so fast, however, that equilibrium is always maintained. The collapse takes only milliseconds, but the time scale of the nuclear reactions is typically from 10^{-15} to 10^{-23} second, so that any departure from equilibrium is immediately corrected.

Another effect was once thought to increase the entropy, but it now seems likely that it actually reduces it somewhat. The high density in the collapsing core favors the reaction known as electron capture. In this process a proton and an electron come together to yield a neutron and a neutrino. The neutrino escapes from the star, carrying off both energy and entropy and cooling the system just as the evaporation of moisture cools the body. There are several complications to this process, so that its effect on the entropy is uncertain. In any case, the loss of the electron diminishes the electron pressure and so allows the implosion to accelerate further.

The first stage in the collapse of a supernova comes to an end when the density of the stellar core reaches a value of about 4×10^{11} grams per cubic centimeter. This is by no means the maximum den-

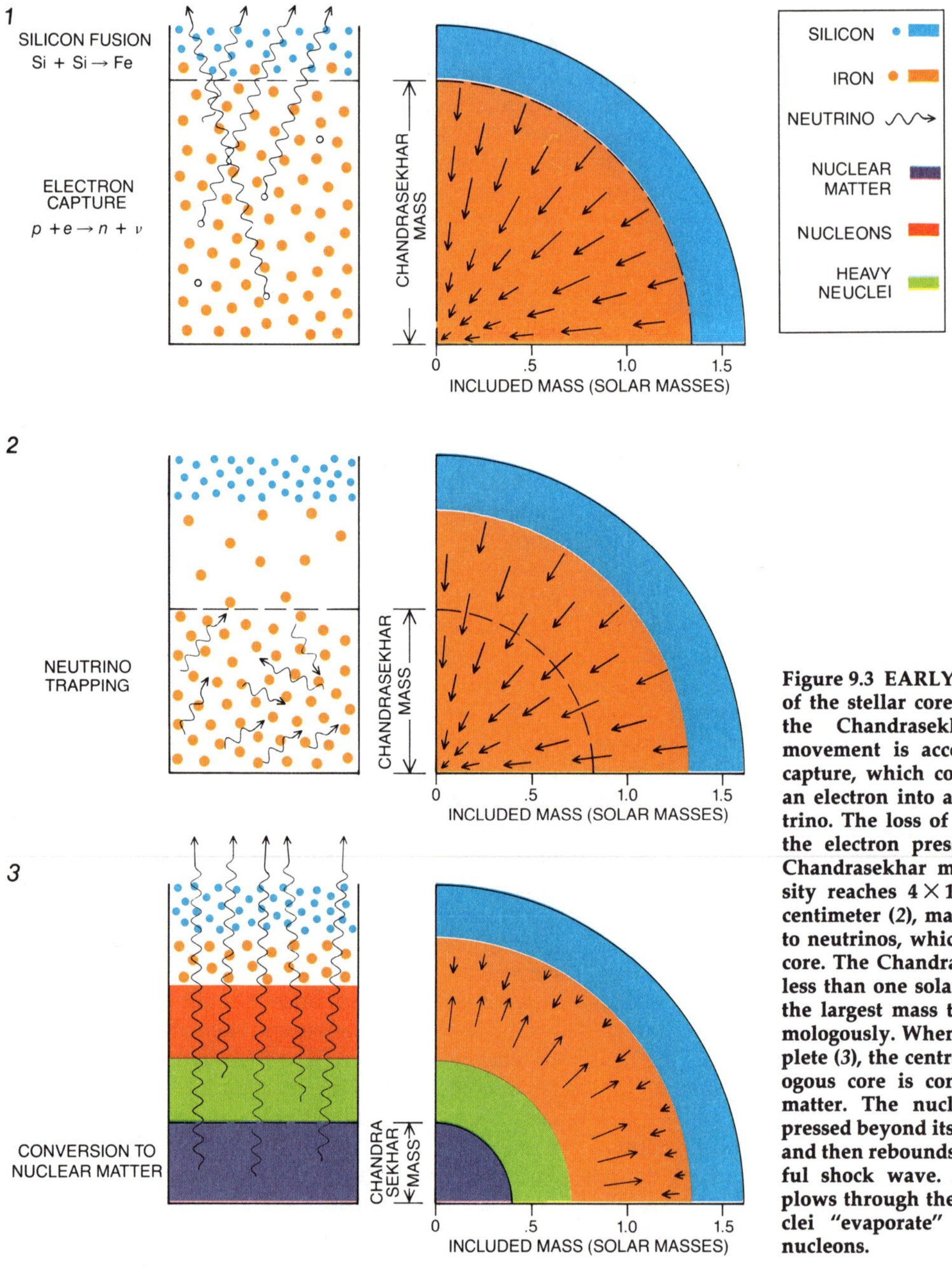

Figure 9.3 EARLY IN THE COLLAPSE of the stellar core (*1*) the iron exceeds the Chandrasekhar mass. Inward movement is accelerated by electron capture, which converts a proton and an electron into a neutron and a neutrino. The loss of the electron reduces the electron pressure and hence the Chandrasekhar mass. When the density reaches 4×10^{11} grams per cubic centimeter (*2*), matter becomes opaque to neutrinos, which are trapped in the core. The Chandrasekhar mass is then less than one solar mass and it is now the largest mass that can collapse homologously. When the collapse is complete (*3*), the central part of the homologous core is converted into nuclear matter. The nuclear matter is compressed beyond its equilibrium density and then rebounds, launching a powerful shock wave. As the shock wave plows through the outer core, iron nuclei "evaporate" to form a gas of nucleons.

sity, since the core continues to contract, but it marks a crucial change in physical properties: at this density matter becomes opaque to neutrinos. The importance of this development was first pointed out by T. J. Mazurek of the Mission Research Laboratory in Santa Barbara, Calif., and by Katsushiko Sato of the University of Tokyo.

The neutrino is an aloof particle that seldom interacts with other forms of matter. Most of the neutrinos that strike the earth, for example, pass all the way through it without once colliding with another particle. When the density exceeds 400 billion grams per cubic centimeter, however, the particles of matter are packed so tightly that even a neutrino is likely to run into one. As a result neutrinos emitted in the collapsing core are effectively trapped there. The trapping is not permanent; after a neutrino has been scattered, absorbed and reemitted many times, it must eventually escape, but the process takes longer than the remaining stages of the collapse. The effective trapping of neutrinos means that no energy can get out of the core.

The process of electron capture in the early part of the collapse reduces not only the electron pressure but also the ratio of electrons to nucleons, the quantity that figures in the calculation of the Chandrasekhar mass. In a typical presupernova core the ratio is between .42 and .46; by the time of neutrino trapping it has fallen to .39. This lower ratio yields a Chandrasekhar mass of .88 solar mass, appreciably less than the original value of between 1.2 and 1.5.

At this point the role of the Chandrasekhar mass in the analysis of the supernova also changes. At the outset it was the largest mass that could be supported by electron pressure; it now becomes the largest mass that can collapse as a unit. Areas within this part of the core can communicate with one another by means of sound waves and pressure waves, so that any variations in density are immediately evened out. As a result the inner part of the core collapses homologously, or all in one piece, preserving its shape.

The theory of homologous collapse was worked out by Peter Goldreich and Steven Weber of Caltech and was further developed by Amos Yahil and James M. Lattimer of the State University of New York at Stony Brook. The shock wave that blows off the outer layers of the star forms at the edge of the homologous core. Before we can give an account of that process, however, we must continue to trace the sequence of events within the core itself.

Chandrasekhar's work showed that electron pressure cannot save the core of a large star from collapse. The only other hope for stopping the contraction is the resistance of nucleons to compression. In the presupernova core nucleon pressure is a negligible fraction of electron pressure. Even at a density of 4×10^{11} grams per cubic centimeter, where neutrino trapping begins, nucleon pressure is insignificant. The reason is the low entropy of the system. At any given temperature, pressure is proportional to the number of particles per unit volume, regardless of the size of the individual particles. An iron nucleus, with 56 nucleons, makes the same contribution to the pressure as an isolated proton does. If the nuclei in the core were broken up, their pressure might be enough to stop the contraction. The fissioning of the nuclei is not possible, however, because the entropy of the core is too low. A supernova core made up of independently moving protons and neutrons would have an entropy per nucleon of between 5 and 8, whereas the actual entropy is less than 1.

The situation does not change, and the collapse is not impeded, until the density in the central part of the core reaches about 2.7×10^{14} grams per cubic centimeter. This is the density of matter inside a large atomic nucleus, and in effect the nucleons in the core merge to form a single gigantic nucleus. A teaspoonful of such matter has about the same mass as all the buildings in Manhattan combined.

Nuclear matter is highly incompressible. Hence once the central part of the core reaches nuclear density there is powerful resistance to further compression. That resistance is the primary source of the shock waves that turn a stellar collapse into a spectacular explosion.

Within the homologously collapsing part of the core, the velocity of the infalling material is directly proportional to distance from the center. (It is just this property that makes the collapse homologous.) Density, on the other hand, decreases with distance from the center, and as a result so does the speed of sound. The radius at which the speed of sound is equal to the infall velocity is called the sonic point, and it marks the boundary of the homologous core (see Figure 9.4). A disturbance inside the core can have no influence beyond this radius. At the sonic point sound waves move outward at the speed of sound, as measured in the coordinate system of the infalling matter. This matter is moving inward at the same speed, however, and so the waves are at a standstill in relation to the center of the star.

When the center of the core reaches nuclear density, it is brought to rest with a jolt. This gives rise to sound waves that propagate back through the medium of the core, rather like the vibrations in the handle of a hammer when it strikes an anvil. The waves slow as they move out through the homologous core, both because the local speed of sound declines and because they are moving upstream against a flow that gets steadily faster (see Figure 9.5). At the sonic point they stop entirely. Meanwhile additional material is falling onto the hard sphere of nuclear matter in the center, generating more waves. For a fraction of a millisecond the waves collect at the sonic point, building up pres-

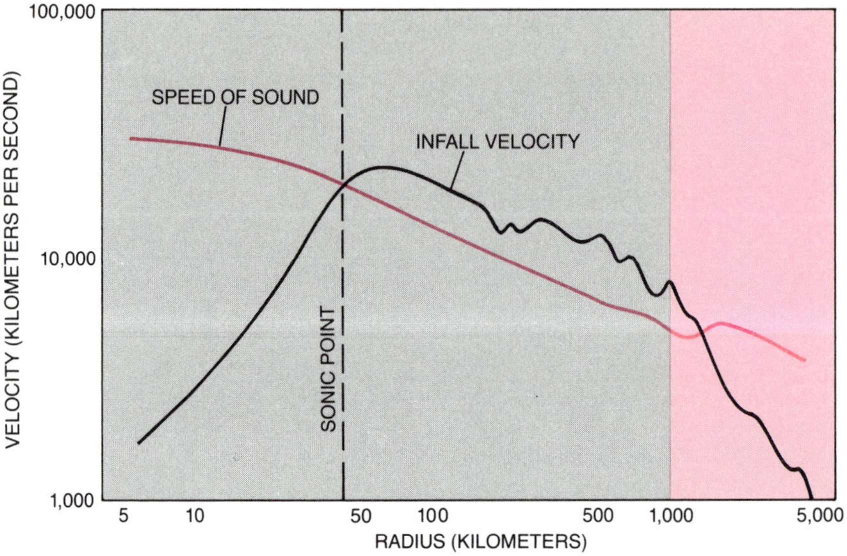

Figure 9.4 SONIC POINT marks the boundary of the homologous core. It is the radius at which the speed of sound is equal to the velocity of the infalling material. A sound wave at the sonic point moves outward at the speed of sound in relation to the material it is passing through, but since that material is falling inward at the same speed, the wave stands still in relation to the center of the star. As a result a disturbance inside the core cannot reach the outside. (Graph based on calculations by W. David Arnett.)

sure there. The bump in pressure slows the material falling through the sonic point, creating a discontinuity in velocity. Such a discontinuous change in velocity constitutes a shock wave.

At the surface of the sphere in the heart of the star infalling material stops suddenly but not instantaneously. The compressibility of nuclear matter is low but not zero, and so momentum carries the collapse beyond the point of equilibrium, compressing the central core to a density even higher than that of an atomic nucleus. We call this point the instant of "maximum scrunch." Most computer simulations suggest the highest density attained is some 50 percent greater than the equilibrium density of a nucleus. After the maximum scrunch the sphere of nuclear matter bounces back, like a rubber ball that has been compressed. The bounce sets off still more sound waves, which join the growing shock wave at the sonic point.

A shock wave differs from a sound wave in two respects. First, a sound wave causes no permanent change in its medium; when the wave has passed, the material is restored to its former state. The passage of a shock wave can induce large changes in density, pressure and entropy. Second, a sound wave—by definition—moves at the speed of sound. A shock wave moves faster, at a speed determined by the energy of the wave. Hence once the pressure discontinuity at the sonic point has built up into a shock wave, it is no longer pinned in place by the infalling matter. The wave can continue outward, into the overlying strata of the star. According to computer simulations, it does so with great speed, between 30,000 and 50,000 kilometers per second.

Figure 9.5 CORE OF A MASSIVE STAR is shown as it passes through the moment of "maximum scrunch," when the center reaches its highest density. Each contour represents a shell of matter whose radial position is followed through a period of 12 milliseconds. The included mass, or total mass inside the contour, does not change as the shells contract and expand. Initially the core is iron, but the extreme compression of the collapse converts the innermost few kilometers into nuclear matter. Surrounding this region is a shell made up of a various heavy nuclei, including iron. At maximum scrunch the contraction stops with a jolt, creating a shock wave (*blue line*) that travels outward. In the wake of the shock nuclei are broken up into individual nucleons.

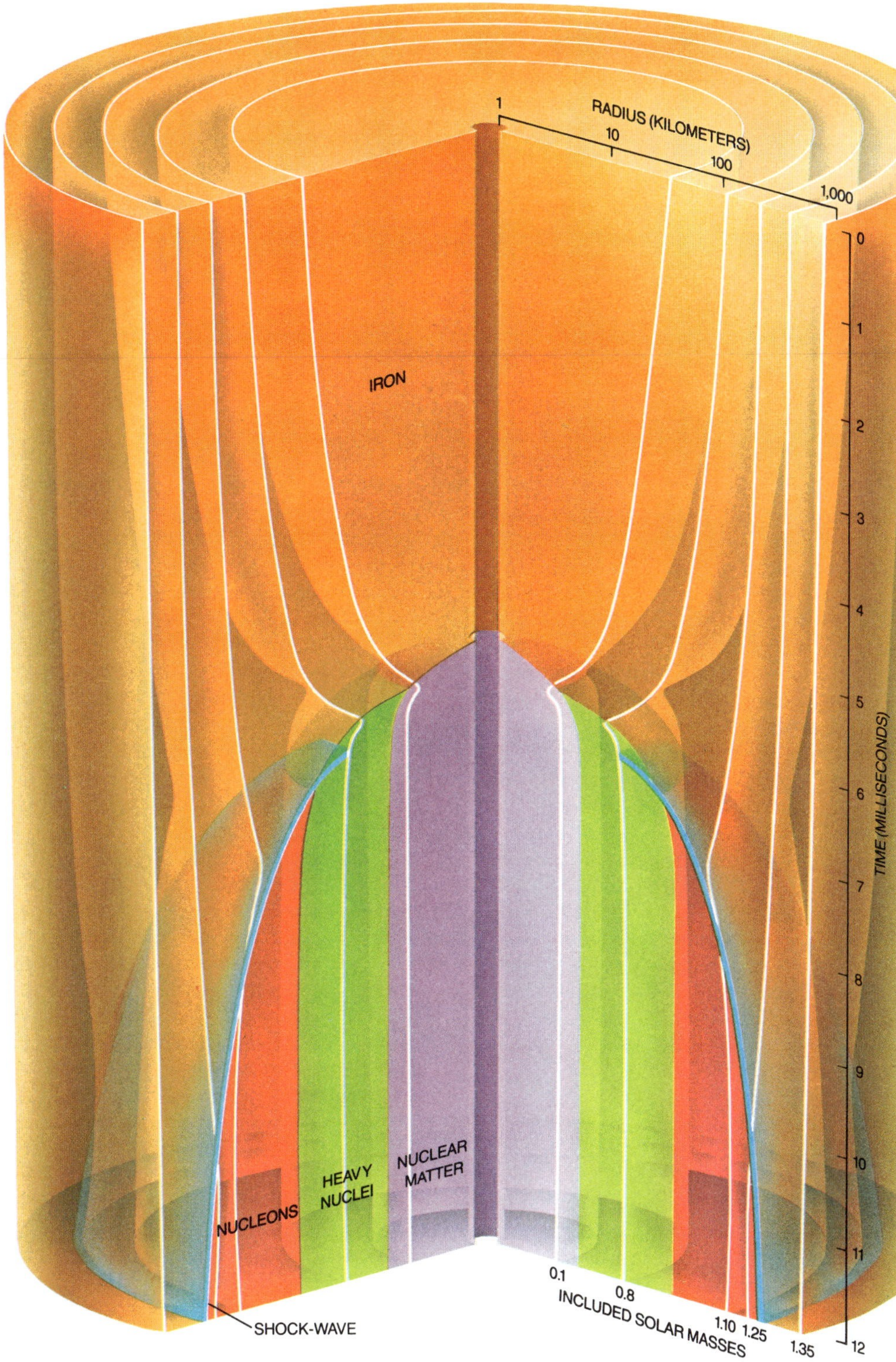

RADIUS (KILOMETERS)
1
10
100
1,000
IRON
TIME (MILLISECONDS)
0
1
2
3
4
5
6
7
8
9
10
11
12
NUCLEONS
HEAVY NUCLEI
NUCLEAR MATTER
SHOCK-WAVE
0.1
0.8
1.10
1.25
1.35
INCLUDED SOLAR MASSES

Up to this point in the progress of the supernova essentially all calculations are in agreement. What happens next, however, is not yet firmly established. In the simplest scenario, which we have favored, the shock wave rushes outward, reaching the surface of the iron core in a fraction of a second and then continuing through the successive onion-like layers of the star. After some days it works its way to the surface and erupts as a violent explosion. Beyond a certain radius—the bifurcation point—all the material of the star is blown off. What is left inside the bifurcation radius condenses into a neutron star.

Alas! Using presupernova cores simulated in 1974 by Weaver and Woosley, calculations of the fate of the shock wave are not so accommodating. The shock travels outward to a distance of between 100 and 200 kilometers from the center of the star, but then it becomes stalled, staying at roughly the same position as matter continues to fall through it. The main reason for the stalling is that the shock breaks up nuclei into individual nucleons. Although this process increases the number of particles, which might be expected to raise the pressure, it also consumes a great deal of energy; the net result is that both temperature and pressure are sharply reduced.

The fragmentation of the nuclei contributes to energy dissipation in another way as well: it releases free protons, which readily capture electrons. The neutrinos emitted in this process can escape, removing their energy from the star. The escape is possible because the shock has entered material whose density is below the critical value for neutrino trapping. The neutrinos that had been trapped behind the shock also stream out, carrying away still more energy. Because of the many hazards to the shock wave in the region between 100 and 200 kilometers, we have named this region of the star the "minefield" (see Figure 9.6).

It would be satisfying to report that we have found a single mechanism capable of explaining for all Type II supernovas how the shock wave makes its way through the minefield. We cannot do so. What we have to offer instead is a set of possible explanations, each of which seems to work for stars in a particular range of masses.

The place to begin is with stars of between 12 and about 18 solar masses. Weaver and Woosley's most recent models of presupernova cores for such stars differ somewhat from those they calculated a decade ago; the most important difference is that the iron core is smaller than earlier estimates indicated—about 1.35 solar masses. The homologous core, at whose surface the shock wave forms, includes .8 solar mass of this material, leaving .55 solar mass of iron outside the sonic point. Since the breaking up of iron nuclei has the highest energy cost, reducing the quantity of iron makes it easier for the shock to break out of the core.

Jerry Cooperstein and Edward A. Baron of Stony Brook have been able to simulate successful supernova explosions in computer calculations that begin with Weaver and Woosley's model cores. The main requirement, first surmised by Sidney H. Kahana of the Brookhaven National Laboratory, is that the homologous core be very strongly compressed, so that it can rebound vigorously and create an intense shock. Two factors cooperate to achieve this result in the simulations. The first factor is the use of general relativity rather than the force field of Newtonian gravitation. The second is the assumption that nuclear matter is much more compressible than had been thought.

Baron's first result showed that a star of 12 solar masses would explode if the compressibility of nuclear matter is 1.5 times the standard value. This seemed rather arbitrary, but then one of us (Brown) examined the problem with a sophisticated method of nuclear-matter theory. It turned out that the most consistent interpretation of the experimental findings yields a compressibility of 2.5 times the standard value! We then found that in 1982 Andrew D. Jackson, E. Krotscheck, D. E. Meltzer and R. A. Smith had reached the same conclusion by another method, but no one had recognized the relevance of their work to the supernova problem. We consider the higher estimate of nuclear compressibility.

The mechanism described by Baron, Cooperstein and Kahana seems to work for stars of up to about 18 solar masses (see Figure 9.7). With still larger stars, however, even the powerful shock wave created in their simulations becomes stalled in the minefield. A star of 25 solar masses has about two solar masses of iron in its core, and so the shock wave must penetrate 1.2 solar masses of iron rather than .55 solar mass. The shock does not have enough energy to dissociate this much iron.

A plausible explanation of what might happen in these massive stars has recently emerged from the work of James R. Wilson of Lawrence Livermore, who has done extensive numerical simulations of supernova explosions. For some time it had seemed

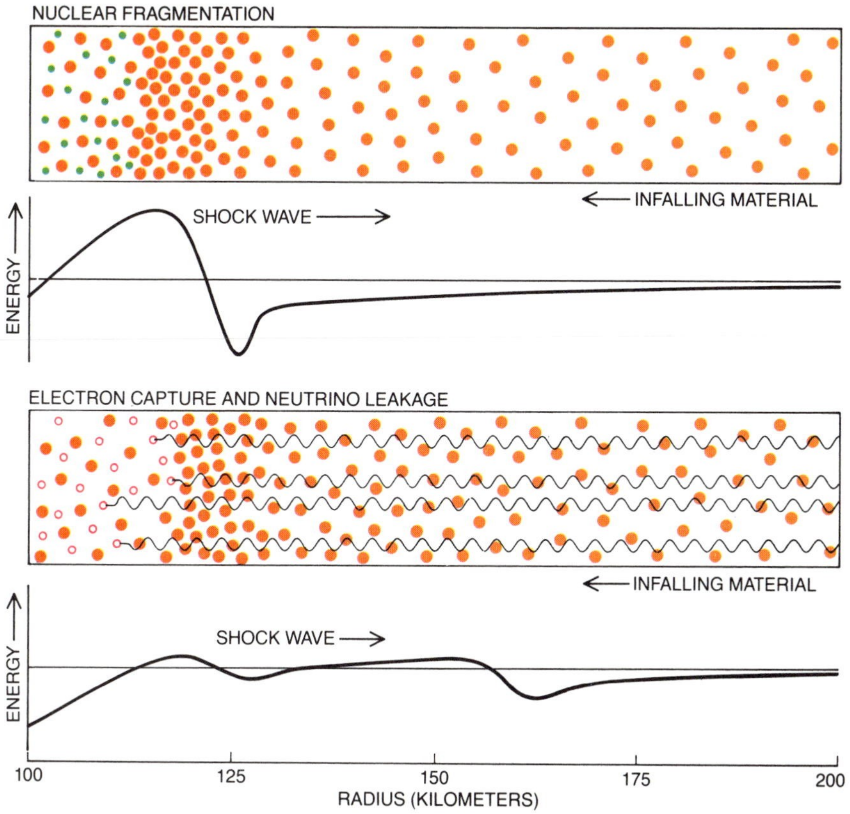

Figure 9.6 SHOCK WAVE SEEMS TO STALL in stars whose mass is greater than about 18 solar masses. Several processes sap the wave's energy. The most important is nuclear fragmentation. Protons released by the fragmentation provide opportunities for electron capture, which further reduces the pressure. Once the wave enters a region of density less than 10^{11} grams per cubic centimeter, leakage of neutrinos carries off more energy. As a result of these effects the shock wave may slow to the speed of the material falling through it and make no further progress. Because of the various hazards to the shock, the authors call the region between 100 and 200 kilometers the "minefield."

that when the shock wave failed, all the mass of the star might fall back into the core, which would evolve into a black hole. That fate is still a possible one, but Wilson noted a new phenomenon when he continued some of his simulations for a longer period.

In the collapsing stellar core it takes only 10 milliseconds or so for the shock wave to reach the minefield and stall. A simulation of the same events, even with the fastest computers, takes at least an hour. Wilson allowed his calculations to run roughly 100 times longer, to simulate a full second of time in the supernova. In almost all cases he found that the shock wave eventually revived.

The revival is due to heating by neutrinos. The inner core is a copious emitter of neutrinos because of continuing electron capture as the matter is compressed to nuclear density. Adam S. Burrows and Lattimer of Stony Brook and Mazurek have shown that half of the electrons in the homologous core are captured within about half a second, and the emitted neutrinos carry off about half of the gravitational energy set free by the collapse, some 10^{53} ergs. Deep within the core the neutrinos make frequent collisions with other particles; indeed, we noted above that they are trapped, in the sense that they cannot escape within the time needed for the homologous collapse. Eventually, though, the neutrinos do percolate upward and reach strata of lower density, where they can move freely.

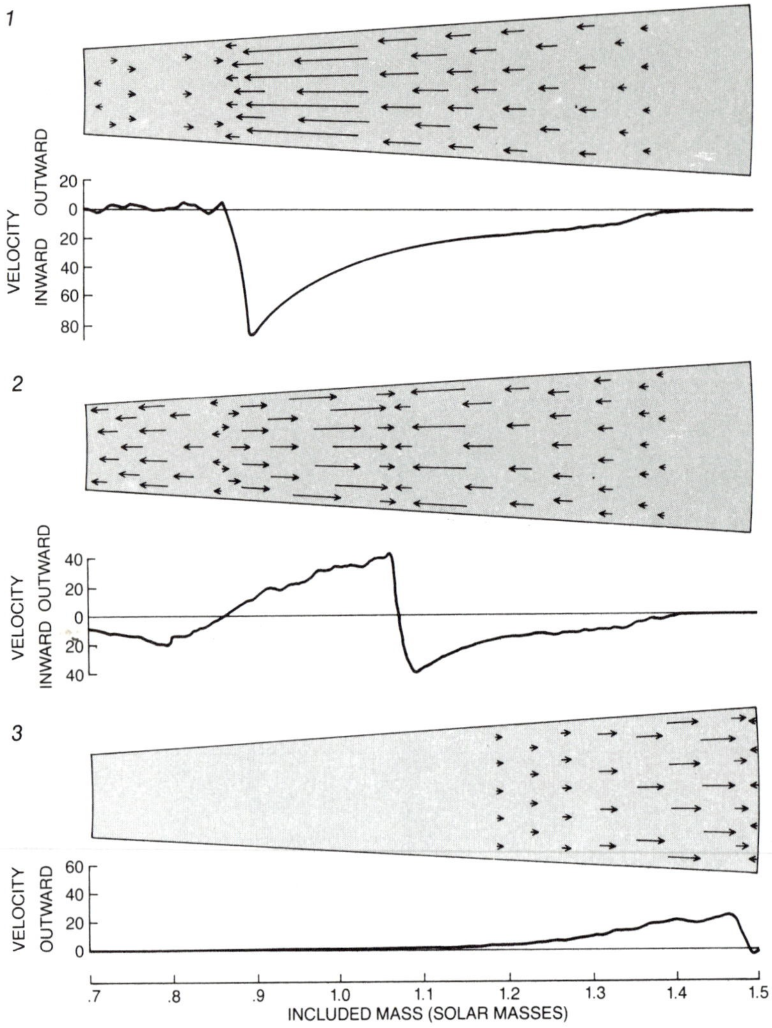

Figure 9.7 SHOCK WAVE can move faster than sound and so it can carry the energy and momentum of the rebound past the sonic point. Just before the bounce (*1*) the inner core has reached the density of nuclear matter and has stopped contracting, but overlying matter is about to fall onto the core at speeds of up to 90,000 kilometers per second. Two milliseconds later (*2*) the core is being driven further inward, but at the same time much of the infalling matter has rebounded to form a shock wave. After 20 milliseconds (*3*) the shock has reached the edge of the core. Velocity profiles were calculated by Jerry Cooperstein. Velocities are given in thousands of kilometers per second.

At the radius where the shock wave stalls only one neutrino out of every 1,000 is likely to collide with a particle of matter, but these collisions nonetheless impart a significant amount of energy. Most of the energy goes into the dissociation of nuclei into nucleons, the very process that caused the shock to stall in the first place. Now, however, the neutrino energy heats the material and therefore raises the pressure sharply. We have named this period, when the shock wave stalls but is then revived by neutrino heating, "the pause that refreshes" (see Figure 9.8).

Neutrino heating is most effective at a radius of about 150 kilometers, where the probability of neutrino absorption is not too low and yet the temperature is not so high that the matter there is itself a significant emitter of neutrinos. The pressure increase at this radius is great enough, after about half a second, to stop the fall of the overlying matter and begin pushing it outward. Hence 150 kilometers becomes the bifurcation radius. All the matter within this boundary ultimately falls into the core; the matter outside, 20 solar masses or more, is expelled.

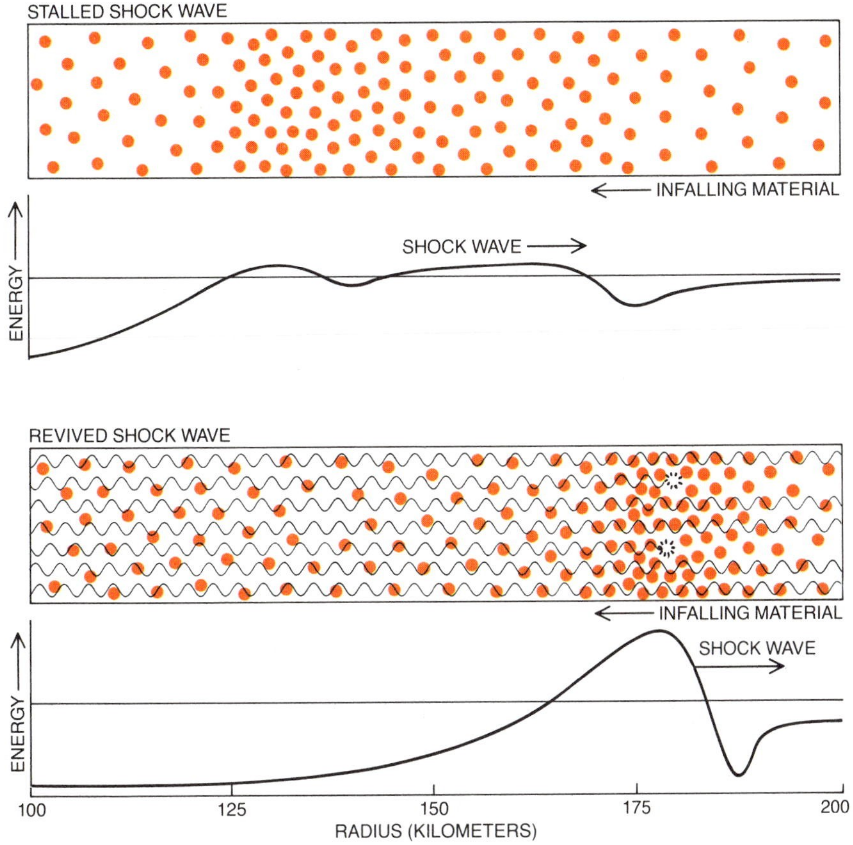

Figure 9.8 REVIVAL OF THE STALLED SHOCK WAVE in heavy stars may be due to heating by neutrinos. Their source is the collapsed core, which radiates the energy equivalent of 10 percent of its mass in the form of neutrinos. Only a small fraction of them are absorbed, but the flux is so intense that many iron nuclei are dissociated. Earlier in the evolution of the supernova the breakup of iron nuclei took energy from the shock wave, but since the process is now powered by external neutrinos, the dissociation no longer decreases shock energy.

The one group of stars left to be considered are those of from eight to 11 solar masses, the smallest stars capable of supporting a Type II supernova explosion. In 1980 Weaver and Woosley suggested that the stars in this group might form a separate class, in which the supernova mechanism is quite different from the mechanism in heavier stars.

According to calculations done by Nomoto and by Weaver and Woosley, in the presupernova evolution of these lighter stars the core does not reach the temperature needed to form iron; instead fusion ends with a mixture of elements between oxygen and silicon. Energy production then stops, and since the mass of the core is greater than the Chandrasekhar limit, the core collapses. The shock wave generated by the collapse may be helped to propagate by two circumstances. First, breaking up oxygen or silicon nuclei robs the shock of less energy than the dissociation of iron nuclei would. Second, farther out in the star the density falls off abruptly (by a factor of roughly 10 billion) at the boundary between the carbon and the helium shells. The shock wave has a much easier time pushing through the lower-density material.

For a star of nine solar masses Nomoto finds that the presupernova core consists of oxygen, neon and magnesium and has a mass of 1.35 solar masses. Nomoto and Wolfgang Hillebrandt of the Max-

Planck Institute for Physics and Astrophysics in Munich have gone on to investigate the further development of this core. They find that the explosion proceeds easily through the core, aided by the burning of oxygen nuclei, and that a rather large amount of energy is released.

Two recent attempts to reproduce the Nomoto-Hillebrandt results have been unsuccessful, and so the status of their model remains unclear. We think the greater compressibility of nuclear matter assumed in the Baron-Cooperstein-Kahana program should be helpful here. Of course it is possible that stars this small do not give rise to supernovas; on the other hand, there are suggestive arguments (based on measurements of the abundance of various nuclear species) that the Crab Nebula was created by the explosion of a star of about nine solar masses.

After the outer layers of a star have been blown off, the fate of the core remains to be decided. Just as gravitation overwhelms electron pressure if the mass exceeds the Chandrasekhar limit, so even nuclear matter cannot resist compression if the gravitational field is strong enough. For a cold neutron star—one that has no source of supporting pressure other than the repulsion of nucleons—the limiting mass is thought to be about 1.8 solar masses. The compact remnant formed by the explosion of lighter stars is well below this limit, and so those supernovas presumably leave behind a stable neutron star. For the larger stars the question is in doubt. In Wilson's calculations any star of more than about 20 solar masses leaves a compact remnant of more than two solar masses. It would appear that the remnant will become a black hole, a region of space where matter has been crushed to infinite density.

Even if the compact remnant ultimately degenerates into a black hole, it begins as a hot neutron star. The central temperature immediately after the explosion is roughly 100 billion degrees Kelvin, which generates enough thermal pressure to support the star even if it is larger than 1.8 solar masses. The hot nuclear matter cools by the emission of neutrinos. The energy they carry off is more than 100 times the energy emitted in the explosion itself: some 3×10^{53} ergs. It is the energy equivalent of 10 percent of the mass of the neutron star.

The detection of neutrinos from a supernova explosion and from the subsequent cooling of the neutron star is one possible way we might get a better grasp of what goes on in these spectacular events. The neutrinos originate in the core of the star and pass almost unhindered through the outer layers, and so they carry evidence of conditions deep inside. Electromagnetic radiation, on the other hand, diffuses slowly through the shells of matter and reveals only what is happening at the surface. Neutrino detectors have recently been set up in mines and tunnels, where they are screened from the background of cosmic rays.

Another observational check on the validity of supernova models is the relative abundances of the chemical elements in the universe. Supernovas are probably the main source of all the elements heavier than carbon, and so the spectrum of elements released in simulated explosions ought to match the observed abundance ratios. Many attempts to reproduce the abundance ratios have failed, but earlier this year Weaver and Woosley completed calculations whose agreement with observation is surprisingly good. They began with Wilson's model for the explosion of a star of 25 solar masses. For almost all the elements and isotopes between carbon and iron their abundance ratios closely match the measured ones.

In recent years the study of supernovas has benefited from a close interaction between analytic theory and computer simulation. The first speculations about supernova mechanisms were put forward decades ago, but they could not be worked out in detail until the computers needed for numerical simulation became available. The results of the computations, on the other hand, cannot be understood except in the context of an analytic model. By continuing this collaboration we should be able to progress from a general grasp of principles and mechanisms to the detailed prediction of astronomical observations.

Young Supernova Remnants

10

The remnants of recent stellar explosions in our galaxy are intense X-ray sources. An orbiting telescope has revealed their structure. One has a pulsar; others are expanding shells of shock-heated gas.

• • •

Frederick D. Seward, Paul Gorenstein and Wallace H. Tucker
August, 1985

The role of chance in scientific discovery is nowhere more apparent than in the study of supernovas. Only five supernovas, or exploding stars, are known to have been observed in our galaxy in the past 1,000 years—the last one in 1604, before the invention of the telescope. The first precise measurements were those of Tycho Brahe, who studied the spectacular "new star" of 1572 for more than a year, gauging its brightness by comparing it with successively dimmer planets and stars until it finally faded from view (see Figure 10.1). Tycho's observations were a crucial episode in the history of astronomy: they led him to break with the Aristotelian tradition, which held that the realm of the "fixed stars" was immutable. Had Tycho's supernova happened a century or so earlier, before the intellectual authority of Aristotelianism had begun to wane, the course of modern astronomy might have been very different. From medieval Chinese records it is known that other explosions were visible on the earth in 1006, 1054 and 1181—the first of these was almost as bright as the half-moon—yet European astronomers largely ignored them.

Modern understanding of supernovas (including the knowledge that they represent the death and not, as Tycho thought, the birth of stars) is based largely on outbursts in distant galaxies. But whereas no one since Johannes Kepler in 1604 has had the chance to study a nearby supernova, workers today can examine the remnants of the explosions observed by Kepler, Tycho and earlier astronomers. Still among the most luminous objects in our galaxy, the remnants consist of hot stellar debris that is hurtling outward from the center of explosion at a speed of roughly 10,000 kilometers per second. The remnants are interesting in their own right, and they may hold clues to the properties of other, more distant energetic objects such as quasars. At the same time they offer a check on theoretical models of the supernova mechanism. By carefully studying a remnant one can derive a rough estimate of the exploded star's mass, which is a critically important variable in all the models.

Such observations are best conducted in the X-ray range of the spectrum. Although supernova remnants are readily detectable with radio telescopes and some are also visible at optical wavelengths, the young ones are so hot that they radiate the bulk of their energy as X rays. The remnant of Tycho's supernova, for example, emits several hundred times

more energy in the X-ray band than the sun does at all wavelengths. Photographed through the largest optical telescopes, however, it is an unprepossessing collection of faint wisps. Even the striking Crab Nebula (as seen on optical photographs), a fossil of the supernova of 1054, radiates most of its energy in the form of X rays.

X rays are absorbed by the earth's atmosphere, and so X-ray astronomers depend on instruments carried aloft on rockets or satellites. The first large orbiting X-ray telescope was the Einstein Observatory, which operated from November, 1978, until April, 1981 [see "The Einstein X-Ray Observatory," by Riccardo Giacconi; Scientific American, February, 1980]. Both in angular resolution and in sensitivity to faint objects the Einstein telescope was a thousandfold improvement over previous instruments. Many of its primary targets were supernova remnants. For the past five years we have been engaged in interpreting the data gathered on these objects.

Long before the X-ray observation of supernova remnants became possible investigators had achieved a general understanding of supernovas. It is widely agreed that a supernova occurs when a star has depleted its nuclear fuel. Throughout its evolution the star maintains its shape by balancing internal pressure against the tendency to collapse under its own gravitation. The internal pressure results from heat generated by nuclear fusion reactions in the core. Initially hydrogen nuclei fuse to form helium; later helium is burned to form carbon. Eventually the fuel in the core is used up, the internal pressure drops and the star contracts.

Most stars do not then explode. A medium-size star (about the size of the sun) collapses until its radius is about equal to the radius of the earth. At that point it is stabilized by a quantum-mechanical effect called degenerate-electron pressure: the electrons resist being crowded too close together. The matter in a collapsed star of this size (called a white dwarf because the collapse has made it white-hot) is a million times denser than ordinary matter.

There is a limit, however, to the amount of gravitation that can be offset by electron degeneracy. In 1935 Subrahmanyan Chandrasekhar showed that a star whose mass is more than approximately 1.4 solar masses will continue to collapse. Ultimately the catastrophic collapse is converted into catastrophic explosion. The manner of the conversion is thought to depend on just how massive the star is; according to current theory, two fundamentally different mechanisms are involved. The mechanisms correspond to the observational classification of supernovas into two main types.

One of the principal bases for classifying a supernova is its light curve: the measure of how its brightness changes with time. Type I outbursts, the rarer of the two classes, all display similar light curves. The brightness of the star increases rapidly during the first few weeks, reaching a peak at which the star may be as luminous as several billion suns, and then falls off gradually over the next six months or so. The light curves of Type II supernovas are more varied, but typically they are about five times fainter at maximum brightness than Type I events, and they fade faster.

Type II supernovas are almost certainly produced by larger stars. Their spectra suggest the explosion takes place in a massive, hydrogen-rich envelope that absorbs much of the radiation. Furthermore, the distribution of Type II supernovas is closely correlated with the distribution of young, bright, massive stars: the outbursts always occur in the arms of spiral galaxies, near star-forming clouds of gas and dust, and they are rarely observed in elliptical galaxies, which consist primarily of old, dim stars. In contrast, Type I supernovas show no preference for spiral arms, and they occur in elliptical galaxies as well. This suggests they are produced by stars that are billions of years old. The mass of such stars cannot be more than a few times that of the sun; otherwise they would have long since consumed their nuclear fuel. (The more massive a star is, the faster it evolves.)

According to the most widely held hypothesis, a Type I supernova arises when a white dwarf has a nearby companion star. If the two are in a close binary orbit, the intense gravitation of the white dwarf can draw matter off the surface of its companion. Eventually the white dwarf, whose

Figure 10.1 TYCHO'S "NEW STAR" appeared in November, 1572, in the constellation Cassiopeia. In this engraving made by the German astronomer Johann Bayer, originally published in 1603, the supernova is the large star under Cassiopeia's right arm. (In Greek mythology Cassiopeia was the wife of King Cepheus and the mother of Andromeda.) When the star appeared, it had a magnitude of −4 and was nearly as bright as Venus. Tycho observed it until it vanished in March, 1574. The remnant of the supernova was discovered 378 years later.

mass is initially less than 1.4 solar masses, accretes enough matter to drive it over the Chandrasekhar limit. At that point it begins to collapse again.

The resulting dramatic increase in temperature and density in the stellar core leads to a new sequence of thermonuclear reactions. Carbon and other light nuclei produced during the normal life of the star fuse to form heavier elements in conjunction with a substantial release of energy. The energy released is enough to disrupt the star completely and to eject the reaction products outward at high velocity. Radioactive nickel decaying to cobalt and iron supplies additional energy to the expanding debris, causing it to glow; many astronomers believe this additional energy explains why the brightness of Type I supernovas declines relatively slowly. The similarity of Type I light curves is attributable to the circumstance that the supernovas all undergo the same radioactive decay processes and are all produced by white dwarfs whose mass is just above the Chandrasekhar limit.

Type II supernovas, on the other hand, are thought to result from the death of stars at least eight times as massive as the sun. The internal temperature of large stars is much greater than that of small ones, and they evolve through a sequence of contractions and fusion reactions that form progressively heavier elements. At the end of the evolution the core consists primarily of iron, the nuclei of which do not fuse spontaneously.

When this inert central mass exceeds a critical limit, the core—but not the gaseous envelope that surrounds it—collapses suddenly. Electrons and protons in the core are compressed to the point where they combine to form neutrons and neutrinos. As the collapse proceeds the degenerate neutrons offer increasing resistance, analogous to the degenerate-electron pressure that stabilizes a white dwarf. When the radius of the core is about 10 kilometers and its density is equivalent to the density inside an atomic nucleus, the collapse comes to a sudden halt. The gravitational energy released by the collapse is carried outward by neutrinos and by shock waves that blow off the envelope (see Chapter 9, "How a Supernova Explodes," by Hans A. Bethe and Gerald Brown).

Although the Type II mechanism is different from the one underlying a Type I supernova, the amount of kinetic energy it imparts to the stellar ejecta is comparable. Whereas a Type I outburst completely disrupts a white dwarf, however, a Type II explosion leaves behind a tiny cinder a billion times denser than a white dwarf: a neutron star.

The theoretical account we have presented so far has gaps. For example, the fate of a star whose mass lies between 1.4 and eight solar masses is unclear. (It is conjectured that the intermediate stars lose enough mass during their lifetime, in the form of stellar wind, to enable them to evolve into white dwarfs.) More fundamental, the entire explanation of Type I and Type II supernovas in terms of the different masses of their stellar progenitors has to be considered unproved: no one has ever observed a star that subsequently exploded. It has been argued, for instance, that at least some Type I supernovas have occurred in regions where stars are being formed, suggesting they were the product of young massive stars rather than of old white dwarfs. In this connection the observation of supernova remnants can be of some help, because the remnants of large stars and of small stars are expected to be different. In particular, the former should include a dense stellar relict.

Such relicts are not always easy to detect. If the original star is very massive, its core will be unstable even as a neutron star and will collapse to a black hole. The gravitational field of a black hole is so strong that no radiation can escape from it; the object is detectable only by its gravitational pull on a companion star, if it has one. A remnant containing an undetected black hole might be mistakenly attributed to the complete disruption of a small star. Likewise a neutron star, which is only 20 kilometers in diameter, might be expected to escape observation from a distance of thousands of light-years.

Fortunately a neutron star, particularly a young one, is extremely energetic. Because a star's magnetic field is tied to the stellar material, the collapse intensifies the field enormously. The magnetic field of a neutron star is estimated to be about 10 trillion times stronger than that of the sun. Furthermore, the collapse increases the star's velocity of rotation by a factor of as much as 100 million because the angular momentum of the star is conserved. The rapidly rotating magnetic field induces a powerful electric field that pulls charged particles off the surface of the neutron star and accelerates them to high speeds. As the particles spiral around the magnetic field lines they emit photons. This effect, known as synchrotron radiation, is strongest near the magnetic poles, which give off intense beams of radiation. Because the star is rotating, the beams are observed on the earth as a regular sequence of pulses, and the neutron star that produces them is called a pulsar.

Most pulsars emit only radio waves. To radiate at

shorter wavelengths the star must rotate rapidly and have an uncommonly powerful magnetic field. Only three pulsars have been found to radiate in the X-ray band. The most luminous one in our galaxy lies in the Crab Nebula, the remnant of the supernova of 1054.

The Einstein X-ray image of the Crab Nebula is dominated by a small region of synchrotron emission—a nebula within the Nebula—around the central pulsar. Ninety-six percent of the X-ray emission comes from electrons moving at velocities close to the speed of light in the synchrotron nebula. The other 4 percent comes from the pulsar itself, which is seen as a bright point source. The spinning star pulses rapidly (30 times per second), but a gradual increase in the pulse period indicates the rotation is slowing down. The star is losing rotational energy at about the same rate as energy is being radiated by the synchrotron nebula in the X-ray, optical and radio bands, which supports the notion that high-energy electrons streaming from the neutron star produce the radiation.

Unlike the Crab Nebula, the remnants of the supernovas of 1006, 1572 (Tycho) and 1604 (Kepler) give no evidence of a pulsar or of any kind of neutron star (see Figure 10.2). If a neutron star were present in a remnant, one would expect to find it on high-resolution X-ray images. At the distance of Tycho's remnant, for example, the thermal radiation from a 400-year-old neutron star should be readily detectable; according to conventional models of the cooling process, such a star would have a surface temperature of several million degrees Celsius. Moreover, although a pulsar would not be visible if the beam did not point toward the earth, a rapidly rotating neutron star should still produce an observable synchrotron nebula. Einstein observations and radio surveys have indeed revealed Crab-like remnants that lack a pulsar, but the remnants of the 1006, 1572 and 1604 supernovas are not among them. It is conceivable that these objects contain unexpectedly cool, slowly spinning neutron stars or even black holes, but we consider it unlikely.

The X-ray images of all three remnants have a shell-like appearance, and all three have been attributed to Type I supernovas. In the case of the outbursts of 1572 and 1604 the evidence for the classification is particularly clear-cut. The observations by Tycho and Kepler were so precise that in the 1940's Walter Baade was able to reconstruct the light curves of the supernovas and show they were typical of Type I events. Chinese records of the outburst of 1006 do not allow a conclusive classification, but its extreme brightness suggests it too was a Type I explosion. On the other hand, some scholars have argued that the Crab supernova of 1054 was a Type II event, although again the recorded observations are inconclusive.

The X-ray observations thus support the idea that there are two distinct mechanisms for producing supernovas, each associated with a particular type of remnant: Type II explosions of massive stars leave Crab-like remnants with neutron stars, and Type I outbursts involving white dwarfs leave hollow spherical shells.

Thanks to its pulsar the Crab Nebula radiates about 70 to 200 times more energy in the X-ray range than the 1006, Tycho and Kepler remnants do. The X rays from the latter objects are not generated by the synchrotron mechanism. Rather they are thermal X rays resulting from collisions of atoms and charged particles in the expanding shell of hot gas. The Crab Nebula probably also has an X-radiating shell that is obscured on the Einstein images by the much brighter emissions from the synchrotron nebula. In the other remnants the large gaseous shell is apparent. It is a tenuous, transparent cloud, yet it may radiate as much energy as a thousand suns.

The shell has two components. One component is debris from the exploded star and the other is interstellar gas swept up by the debris. Initially the fastest stellar ejecta travel outward from the explosion with velocities of between 10,000 and 20,000 kilometers per second. Interstellar gas, however, forms a barrier that offers increasing resistance to the expansion. As the leading edge of the shell plows into the interstellar material, two shock waves form. One wave moves ahead of the ejecta and the other, called the reverse shock, moves (with respect to the first shock) back into the ejecta (see Figure 10.3). Seen from the outside both waves move outward. The ejecta tend to become unstable and break into clumps as they push into the interstellar gas, which is itself not distributed uniformly. As a result the shock fronts do not propagate with the same velocity in all directions, and so they are not perfectly spherical.

The hottest material lies between the two shock waves. Here the previously cool interstellar gas has been heated and compressed by the expanding ejecta, which have in turn been slowed and heated by the collision. Both the shocked ejecta and the shocked interstellar gas radiate in the X-ray range.

The spectra of the shell-like remnants prove that

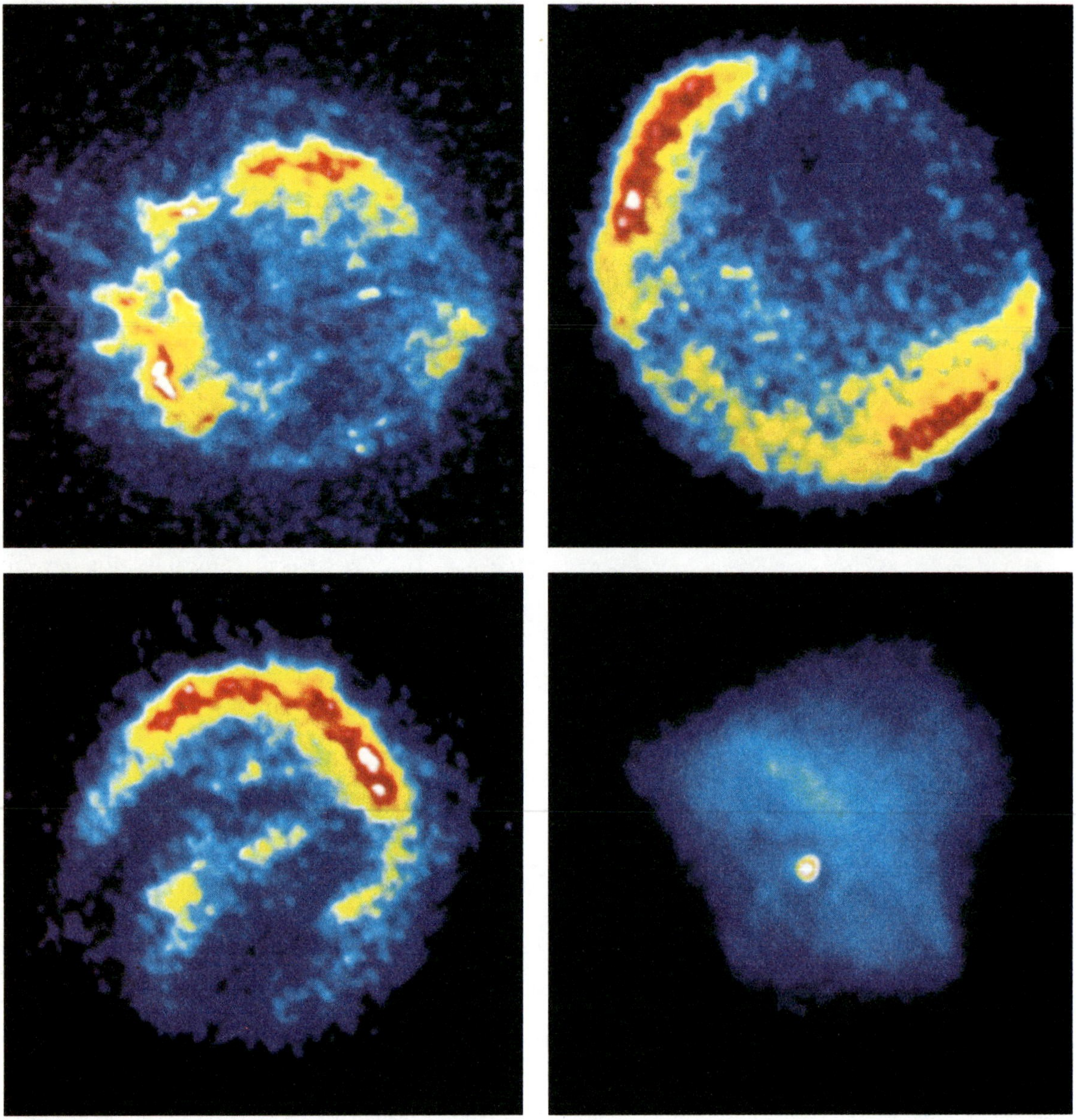

Figure 10.2 TWO TYPES OF SUPERNOVA REMNANT are illustrated by X-ray images from the Einstein Observatory. Like Tycho's remnant, Cassiopeia A (*upper left*), SNR 1006 (*upper right*) and Kepler's remnant (*lower left*) are shell-like: their X rays are thermal emissions from an expanding shell of stellar ejecta and swept-up interstellar gas. None of them seems to incorporate a central pulsar. In contrast, the X rays from the Crab Nebula (*lower right*) are synchrotron radiation from a small region around a bright central pulsar. The glare of the synchrotron nebula may obscure a surrounding shell; the Crab Nebula is 170 times as bright as SNR 1006 even though it is about the same age. (The true relative sizes and brightnesses of the remnant are given in Figure 10.6.)

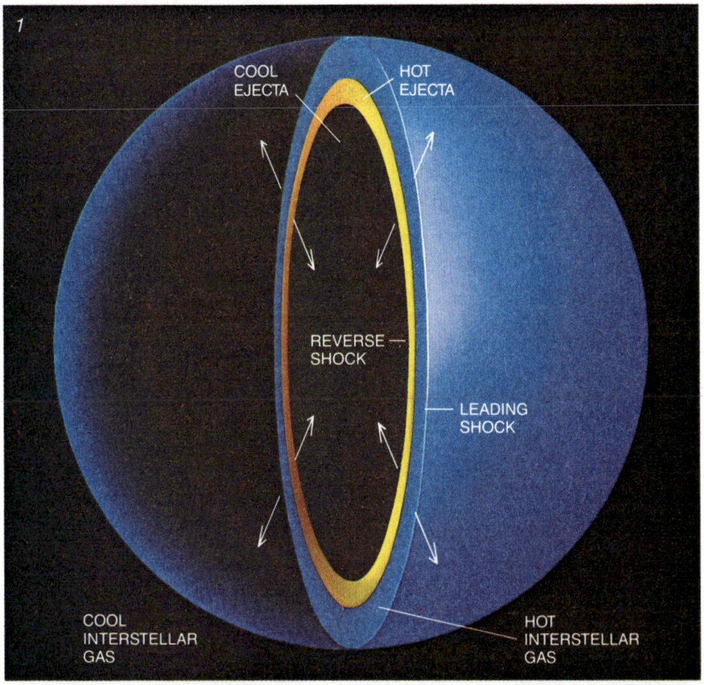

Figure 10.3 SHOCK WAVES generated by a supernova explosion account for the X-ray and radio emissions from a shell-like remnant. As the expanding shell pushes into the interstellar medium a shock wave precedes it, heating the interstellar gas. The collision slows the expansion. A reverse shock front forms, moving inward with respect to the first shock and heating the ejecta. The hot ejecta, and to a lesser extent the interstellar gas, emit thermal X rays. Shock-accelerated electrons emit radio waves by the synchrotron process. In an idealized supernova explosion (*1*) the shock fronts are perfectly smooth and spherical. In reality (*2*) the shock wave in the interstellar gas is indeed relatively smooth (see Figure 10.5), but the reverse shock wave is not because the unevenly distributed ejecta break into clumps. Seen in projection (*3*), the shell of clumpy ejecta produces an incomplete ring.

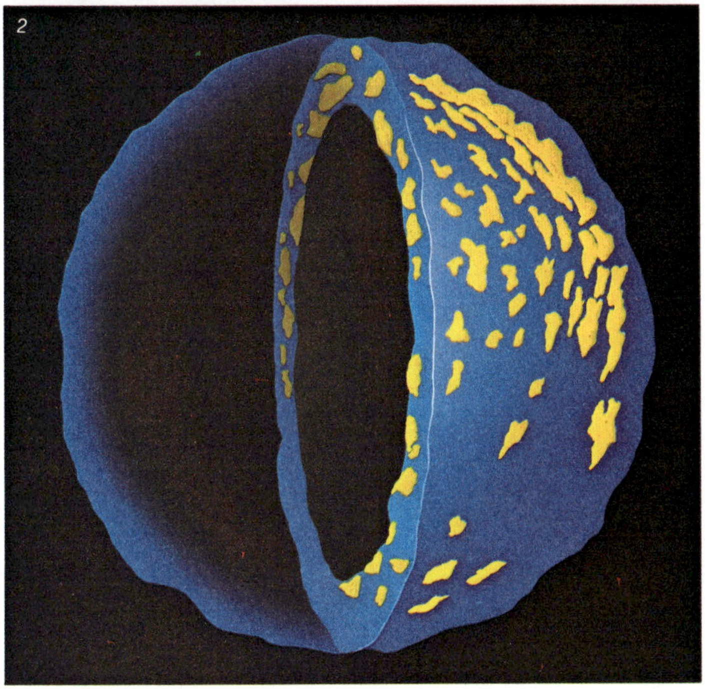

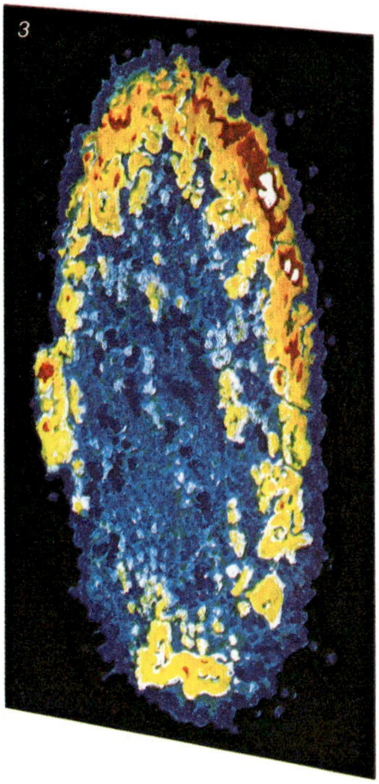

the X-radiation emanating from them is indeed of thermal origin (see Figure 10.4). Most of their energy is radiated at discrete frequencies in the form of atomic emission lines. The lines are produced in a hot gas by atoms or ions that have been excited by collisions with charged particles. Such a collision raises one of the atom's electrons to a higher energy level; the electron then falls back to a lower energy level by emitting a photon. Each emission line is associated with a particular energy transition and hence reveals the presence in the gas of atoms of a particular substance. Synchrotron radiation, in contrast, is emitted across a continuous range of frequencies by free electrons in a magnetic field, and so it reveals nothing about the composition of the gas cloud.

From the strength of their emission lines it is apparent that the levels of silicon, sulfur, argon and calcium are significantly higher in the shell-like remnants than they are in the sun or in the interstellar gas. Tycho's remnant is enriched in these "medium weight" elements by about a factor of six. The enrichment can only be attributed to the supernova. Thus the X-ray spectra provide convincing evidence for a theory frequently stated but seldom directly verified: that the elements heavier than helium are produced inside stars and are scattered through space by supernova explosions. (As the

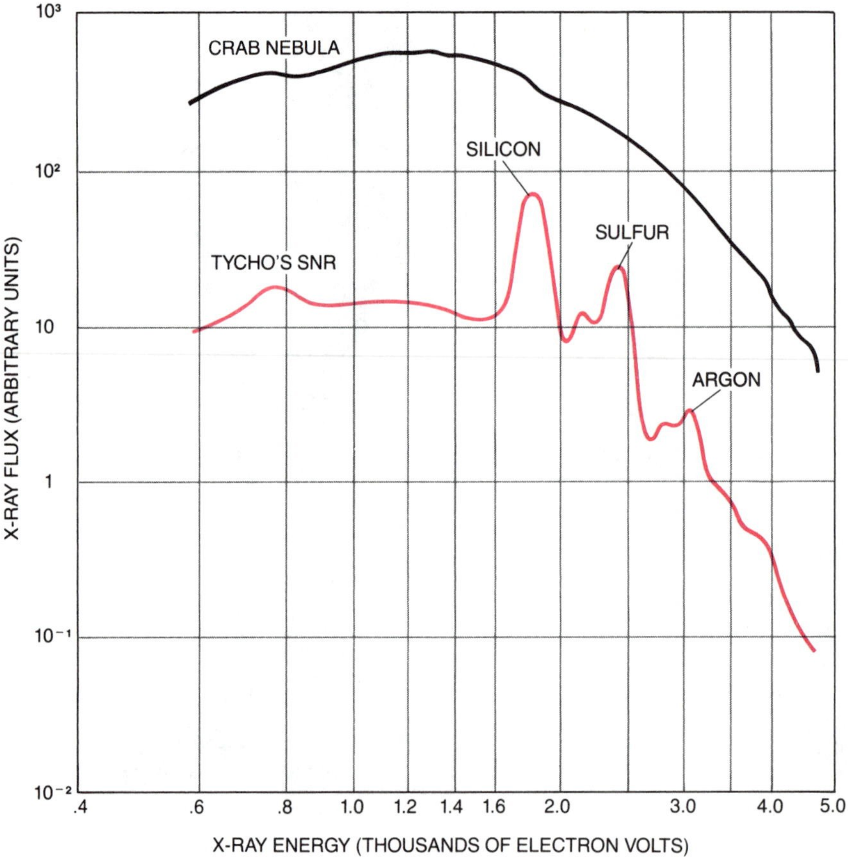

Figure 10.4 X-RAY SPECTRA of the Crab Nebula and of Tycho's supernova remnant. The Crab Nebula contains a pulsar that emits electrons moving at speeds close to the speed of light. As the electrons spiral around the magnetic field lines in the nebula, they emit synchrotron radiation in a smooth continuum across the entire X-ray band. In contrast the spectrum of Tycho's remnant shows strong thermal emission lines from ions of silicon, sulfur and argon formed during the supernova explosion. The ions are excited by collisions in the gas, whose temperature is roughly six million degrees Celsius. (Spectra prepared by the Goddard Space Flight Center using the Einstein Observatory solid-state spectrometer.)

shells expand over hundreds of thousands of years, the heavy elements become dispersed in the interstellar medium.) One puzzling feature of the spectra, however, is the lack of evidence for elements heavier than calcium—specifically, for the nickel, cobalt and iron expected to be manufactured in Type I explosions. Perhaps these iron-group elements are present at too low a temperature to radiate in the X-ray band.

In principle the X-ray observations enable one to calculate the mass as well as the composition of a thermally radiating remnant. A fairly simple formula relates the luminosity of a gas cloud to its temperature, volume and density. The temperature can be inferred from the spectrum of the gas, and so one can derive the density of a remnant from its observed luminosity. If the X-ray data make it possible to distinguish between stellar ejecta and swept-up interstellar gas in the shell, the mass of each can be computed separately. Because the mass of the ejecta should approximately equal the mass of the exploded star, the computation indicates whether the star could have been a white dwarf, as theory predicts for Type I supernovas.

We have done the calculation for Tycho's remnant. On its X-ray image the shocked interstellar gas and the ejecta are distinguishable (see Figure 10.5). The interstellar gas is faintly visible at the edge of the image, just outside the main ring where the emissions are brightest. The ring is probably formed by clumps of ejecta superposed along the line of sight; individual clumps are clearly resolved in the center of the image. We calculated the average size and mass of the central clumps. By assuming the averages also apply to the unresolved clumps in the ring, we determined that the shell of Tycho's remnant contains some 400 clumps amounting to about one solar mass of material. Diffuse emission from the shell indicates the presence of another solar mass of unaggregated debris, making the total mass of the ejecta about twice the mass of the sun.

That is a bit high for a white dwarf, which would have no more than 1.4 solar masses. The discrepancy may be attributable to the uncertainties in our calculation. In particular the distance of Tycho's remnant (or of any of the other remnants) is not known precisely; we assumed a distance of 10,000 light-years. If the real distance is 15 percent less, the remnant's intrinsic X-ray luminosity would be smaller and the estimated mass of its stellar progenitor would fall to about 1.5 solar masses. Our results are therefore compatible with the theory that Tycho's remnant was produced by the complete disruption of a white dwarf. Although we cannot unequivocally rule out the possibility that cool, unobservable ejecta, or even a neutron star, are hidden inside the hot shell, we think the weight of the evidence favors a low-mass explosion.

From the brightness of the shocked interstellar gas we have also estimated the mass of the swept-up material. It too is about two solar masses. Tycho's remnant appears to be in an intermediate stage of development in which the swept-up material is beginning to have a significant effect on the shell. In a few more centuries interstellar material will dominate the dynamics of the shell's expansion and will account for most of the radiation. As the material becomes thoroughly mixed with the ejecta, most of the information on the supernova explosion—the initial expansion velocity and the mass and heavy-element content of the star—will be lost. The mixing of ejecta and interstellar gas already makes it difficult to estimate the mass of the supernova of 1006. The remnant of the 1006 event, judging from its X-ray spectrum, is less enriched in the heavy elements than Tycho's remnant is, indicating that its stellar material has been more diluted.

The remnant of Kepler's supernova of 1604 presents another problem: it is probably twice as far away as Tycho's remnant, and on the Einstein image the shocked interstellar gas is not easily distinguishable from the ejecta. It might be possible anyway to estimate roughly the mass of the star, but we have not tried to do so. As we stated above, however, the absence of evidence for neutron stars in the 1006 and Kepler remnants supports the hypothesis that they were formed by exploding white dwarfs.

The absence of a neutron star in another shell-like remnant, one we have not mentioned so far, is more troubling. Called Cassiopeia A, it is the youngest remnant; the expansion velocity of the shell implies that the supernova occurred in the last decades of the 17th century. There is no conclusive evidence that the outburst was witnessed at all on the earth (although William B. Ashworth, Jr., of the University of Missouri has argued that John Flamsteed, the first astronomer royal of England, observed a "new star" in approximately the right location in 1680). Since the supernova did not arouse widespread interest, it certainly could not have been very bright, and this strongly suggests it was a Type II event. According to theory, the remnant should have a dense star at its center, but the X-ray image does not reveal the synchrotron nebula one would expect

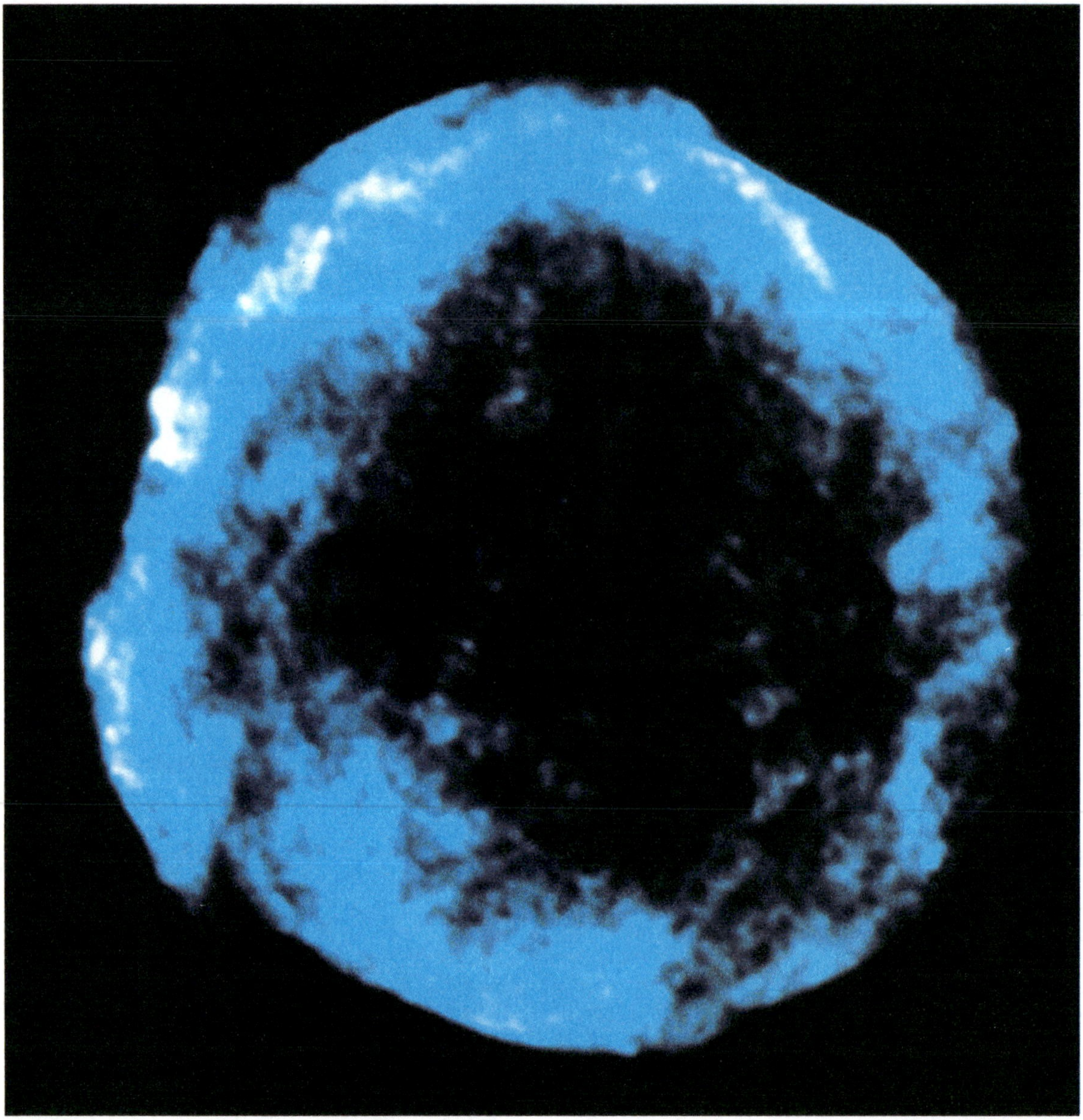

Figure 10.5 RADIO AND X-RAY IMAGES of Tycho's supernova remnant are strikingly similar. At both radio (*left*) and X-ray (*right*) wavelengths the bulk of the emissions come from a shell whose thickness is roughly 30 percent of the remnant's radius. On the X-ray image the clumps of hot stellar ejecta in the shell account for the brightest emissions (*yellow, red and white*); individual clumps are resolved near the center of the image. Swept-up interstel-

around a neutron star. In the absence of observational evidence one can only speculate that Cassiopeia A may contain a black hole.

In addition to being a luminous X-ray object Cassiopeia A is the brightest source of radio emissions in the sky; the other shell-like remnants can also be mapped in detail at radio wavelengths. (Cassiopeia A, Tycho's remnant and many older remnants were actually discovered as radio objects.) Whereas the X rays from these remnants are thermal, the radio emissions are not. They are synchro-

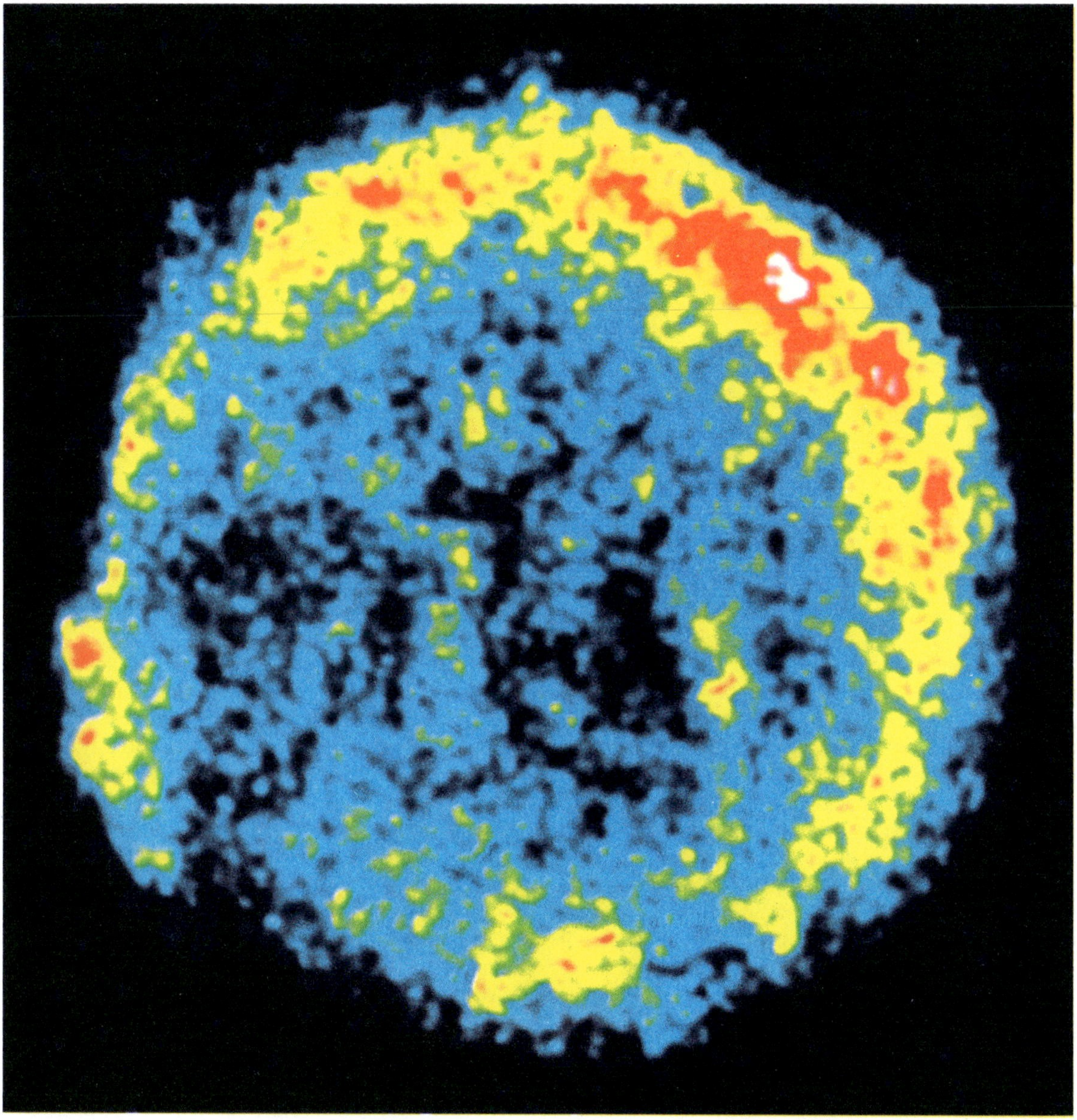

lar gas forms a layer of faint emission outside the shell. This layer is blue on the X-ray image and is most clearly visible at the upper-right edge. The radio map was made at a wavelength of 20 centimeters by David Green and Steven Gull with the five-kilometer telescope of the University of Cambridge. The X-ray image was prepared by the authors from data gathered with the high-resolution imager on the orbiting Einstein Observatory.

tron radiation produced by enormous numbers of highly energetic electrons moving in a magnetic field. The electrons cannot be as energetic as the ones in the Crab Nebula or they too would emit X rays. Nevertheless, they do achieve velocities approaching the speed of light. Given that the shell-like remnants appear to contain no neutron stars, where do the energetic particles come from?

The question is of particular interest because a similar phenomenon seems to operate on a much larger larger scale in active radio galaxies and in quasars. In the case of supernovas the initial explo-

REMNANT	SUPERNOVA YEAR	SUPERNOVA TYPE	DISTANCE (LIGHT-YEARS)	ANGULAR DIAMETER OF X-RAY EMISSION (MINUTES OF ARC; FULL MOON = 30)	INTRINSIC DIAMETER OF X-RAY EMISSION (LIGHT-YEARS)	RELATIVE X-RAY INTENSITY
SNR 1006	1006	I?	3,300	30	28	6
CRAB NEBULA	1054	II?	6,500	1.5 × 1.7	2.8 × 3.2	1,000
SNR 1181	1181	II?	8,500	4 × 6	10 × 15	0.2
TYCHO'S SNR	1572	I	10,000	8	22	14
KEPLER'S SNR	1604	I	16,300 – 32,600	3	14 – 28	5
CASSIOPEIA A	~1680	II?	9,100	4	10	40

Figure 10.6 SIX SUPERNOVA REMNANTS in our galaxy are associated with stellar explosions of the past 1,000 years. None of the distance and size estimates are precise, but the estimates for Kepler's remnant are particularly uncertain. The date of the supernova associated with Cassiopeia A is unclear because there is no proof it was seen. The supernovas of 1006, 1054 and 1181 were recorded in China.

sion cannot account for the presence in the remnant of high-energy electrons, because over the centuries the particles either would have escaped from the remnant or would have radiated most of their energy. Somehow the shock waves from the explosion must continue to accelerate electrons to high energies.

If the shock waves do account for the radio emissions, one would expect the radio and X-ray images of a shell-like remnant to be similar because most of the X rays come from shock-heated matter. Indeed, the correspondence between the two images is striking in all the young remnants we have examined. Combined radio and X-ray observations may in the future help to reveal the details of the shock-acceleration mechanism. According to one promising hypothesis, the ionized clumps of gas in the shocked ejecta form turbulent magnetic eddies; electrons in the remnant are accelerated by collisions with the clumps and emit radio waves as they spiral around the magnetic field lines.

The discovery of Tycho's remnant with the Jodrell Bank radio telescope in 1952, some 378 years after the star had disappeared from view, was a milestone in astronomy. It was altogether appropriate that the value of the new technology should be demonstrated on an object associated with Tycho, since he was the first of a long line of astronomers who have devoted much of their energy—of which the flamboyant Tycho had a great deal—to the improvement of astronomical instruments. In the decades since the Jodrell Bank discovery X-ray images of remnants have added still another dimension to the study of supernovas.

Yet nothing learned from the observation of remnants matches what astronomers could learn about supernovas if they had the chance to see another bright explosion in our galaxy. Because nearby stars are by now well catalogued, it would be possible, for instance, to say whether the star that exploded was a white dwarf or a more massive object. When the next galactic supernova occurs, astronomers will be ready, thanks to the tradition begun by Tycho, with an arsenal of powerful instruments. Their observations will no doubt open up intriguing new avenues of investigation.

The Great Supernova of 1987

11

On February 23 of that year astronomers gained their first closeup view of a star's cataclysmic death since 1604. Worldwide observations have tested existing theory and added new puzzles.

• • •

Stan Woosley and Tom Weaver
August, 1989

The collapse and explosion of a massive star is one of nature's grandest spectacles. For sheer power nothing can match it. During the supernova's first 10 seconds, as the star's core implodes to form a neutron star, it radiates as much energy from a central region 20 miles across as all the other stars and galaxies in the rest of the visible universe combined. To put it another way, the energy of that 10-second burst is 100 times more than the sun will radiate in its entire 10-billion-year lifetime. It is a feat that stretches even the well-stretched minds of astronomers.

Yet supernovas are more than distant spectacles: they make and expel the seeds of life. Only the simplest and lightest elements, hydrogen and helium, were formed in the primordial fireball of the big bang. Most of the heavier elements, including the carbon of our chemistry, the iron in our blood and the oxygen we breathe, were forged in supernovas long before the solar system took shape.

Important as they are, few supernovas have been seen nearby. The last one in our own galaxy flared in 1604, shortly before the invention of the telescope; Johannes Kepler, who observed it, was able to record only its brightness and duration. In the absence of nearby events, understanding of many features of supernovas has remained largely theoretical. Telescopes do reveal a dozen or so events each year in distant galaxies, and careful study of a few distant supernovas has served for testing some coarser aspects of theory. But none was close enough for the modern panoply of ground- and space-based instruments to chronicle the event in detail.

All that changed on the night of February 23, 1987, when a burst of light and a pulse of the elusive particles called neutrinos reached the earth from the brightest supernova in 383 years. Light from the explosion, 160,000 light-years away in the Large Magellanic Cloud, a satellite galaxy of our own, was visible only in the Southern Hemisphere. It is a tribute to the care with which amateur and professional observers monitor the southern sky that the supernova was photographed within an hour of the time its first light must have arrived—although the observer, Robert McNaught of Siding Spring, Australia, did not realize he had captured it until later.

About 20 hours after McNaught's first photograph, Ian Shelton of the Las Campanas Observatory in Chile was photographing the Large Magellanic Cloud. Comparing a photograph made that night with one from the night before, he found a new, starlike image on the later plate. The image was very bright—so bright that it ought to be visible to the naked eye. Shelton walked outside and looked up. Supernova 1987A (*A* for the first supernova, bright or faint, to be found that year) had been discovered.

Within a day anyone who had any astronomical instrument in the Southern Hemisphere was marveling at the sight. During the following months the array of instruments trained on the supernova came to include telescopes and sensors on board balloons, rockets, satellites and an airplane, as well as ground-based instruments of all descriptions. By now, more than two years later, the supernova has been studied at all wavelengths of the electromagnetic spectrum, and it is the first astronomical source of neutrinos to have been detected other than the sun. Together the observations give a coherent picture of the grand event (see Figure 11.1), a picture that vindicates theory but also holds some surprises.

A supernova's characteristics are shaped by the progenitor star. In the broadest terms, SN 1987A is a type II supernova, powered by the gravitational collapse of a stellar core—a catastrophe unique to massive stars. (Type I supernovas, which include the 1604 event, are thought to be thermonuclear explosions of white-dwarf stars to which a critical mass of material has been added.) To make sense of what was observed in SN 1987A, it is best to begin with the history of the star that exploded. The story that follows is based on computer simulations of the evolution of a hypothetical massive star, which we and others (including Ken'ichi Nomoto and his colleagues at the University of Tokyo and W. David Arnett of the University of Arizona) have

Figure 11.1 AGED STAR AND ITS BRILLIANT DEATH are seen in photographs of the same region of the Large Magellanic Cloud made a few months apart. The progenitor star (*inset*), a blue supergiant called Sanduleak −69° 202, was about 80,000 times brighter than the sun; at its brightest (in May, 1987), the supernova reached 200 million solar luminosities. Even so, light represented only a tiny fraction of the total output of supernova 1987A: 30,000 times more energy was discharged in a burst of elusive particles called neutrinos.

developed over the past 25 years in an effort to understand type II events. Since the supernova—the first to occur in an identified star—we have recalculated our model to take into account the special features of the star known beforehand as Sanduleak −69°202, after the astronomer Nicholas Sanduleak, who catalogued it about 20 years ago.

The story begins about 11 million years ago in a gas-rich region of the Large Magellanic Cloud known as 30 Doradus, or the Tarantula Nebula, where a star was born with about 18 times the mass of the sun. For the next 10 million years this star, like most others, generated energy by fusing hydrogen into helium. Because of its great mass the star had to maintain high temperatures and pressures in its core to avoid collapse; as a result it was much more luminous than the sun—about 40,000 times as bright—and a profligate burner of nuclear fuel.

When hydrogen had finished fusing into helium in the innermost 30 percent of the star, the central regions began a gradual contraction. As the core was compressed over tens of thousands of years, from a density of six grams per cubic centimeter to 1,100, it heated up from about 40 million degrees Kelvin to 190 million degrees. The higher core temperature and pressure ignited a new and heavier nuclear fuel, helium. At the same time the outer layers of the star (mostly unburned hydrogen) responded to the additional radiation from the hotter core by expanding to a radius of about 300 million kilometers, or about twice the distance from here to the sun. The star had become a red supergiant.

The core's supply of helium was exhausted in less than a million years, burned to carbon and oxygen. During the few thousand years that remained to the star, this scenario of core contraction, heating and ignition of a new and heavier nuclear fuel—the ash of a previous cycle of fusion—was played out repeatedly. Carbon was the next to burn, at a core temperature of 740 million degrees K. and a density of 240,000 grams per cubic centimeter, yielding a mixture of neon, magnesium and sodium. Then came neon, at 1.6 billion degrees and 7.4 million grams per cubic centimeter, followed by oxygen (2.1 billion degrees and 16 million grams per cubic centimeter) and finally silicon and sulfur (3.4 billion degrees and 50 million grams per cubic centimeter). Because ignition of successively heavier fuels took place in the very center of the star while previous fuels continued to burn in the less dense, overlying regions, the interior of the star came to resemble a cosmic onion, with elements layered in order of increasing atomic weight toward the center (see Figure 11.2).

The core of the star passed through consecutive stages of burning at an accelerating pace. Whereas the burning of helium had lasted nearly a million years, carbon took 12,000 years, neon perhaps 12

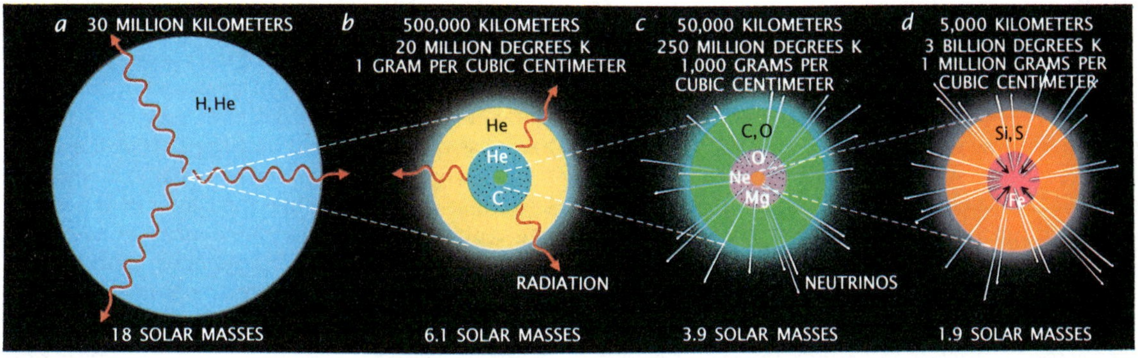

Figure 11.2 COSMIC ONION—the structure of the presupernova star in its final moments—is made up of concentric shells of successively heavier elements undergoing nuclear fusion. The radius of each shell, the temperature and density at its surface and the mass the shell includes are given; stippling indicates regions undergoing convection. When the center of the star's vast, tenuous envelope of hydrogen and helium (*a*) is magnified by a factor of 30 (*b*), a core of helium four times the diameter of Jupiter is revealed. Enlargement by another factor of 10 (*c*) exposes the ash from helium-burning: a core of carbon and oxygen, (*d*). A final step of fusion has turned silicon and sulfur into 1.4 solar masses of iron at the very center of the star.

years, oxygen four years and silicon, at the end, just a week. Each stock of nuclear fuel after hydrogen released about the same total energy, but at core temperatures above 500 million degrees K., beginning with carbon-burning, the star found a new and far more efficient way to spend its energy capital. Very energetic gamma-ray photons, abundant at such temperatures, were transformed into particle pairs—an electron and a positron, an electron's anti-matter counterpart—as they passed near atomic nuclei. The particles promptly annihilated each other, usually recreating the gamma rays but sometimes giving rise to neutrinos.

Neutrinos hardly interact with matter at all. They escaped from the star far more easily than the original gamma rays could have, carrying off energy. Even during carbon-burning, neutrino energy loss exceeded the energy loss by radiation. As the core's temperature rose during the later stages of its evolution, the neutrino luminosity rose exponentially to become a ruinous energy drain, hastening the star's demise.

This late evolution of the core proceeded too fast to have any effect on the star's vast envelope of hydrogen. Yet it turned out that the envelope had also evolved since the star had become a red supergiant. When workers first determined which star had exploded, they were surprised to find the progenitor star was not a red supergiant, as most stellar-evolution models for type II supernovas had predicted, but a blue supergiant—a smaller and hotter star [see "Helium-rich Supernovas," by J. Craig Wheeler and Robert P. Harkness; SCIENTIFIC AMERICAN, November, 1987].

The star's envelope, and not just its core, had apparently contracted beginning perhaps 40,000 years before the explosion, after the helium that had powered its red-supergiant stage was exhausted (see Figure 11.3). Theorists are still debating the reasons, but the distinctive composition of star-forming gas in the Large Magellanic Cloud may have been the most important factor: in comparison with our own galaxy, the gas has a much lower content of elements heavier than helium. Among those elements, oxygen plays a special role in the evolution of a star. A lower oxygen content makes a star's envelope more transparent to radiation and hence perhaps more likely to contract. Oxygen also serves as a catalyst in the generation of energy by hydrogen fusion. Modeling suggests that a low initial oxygen content might subtly modify the early evolution of a massive star so as to ultimately yield a blue, rather than a red, supergiant.

The small radius of the progenitor star was to have dramatic effects later, when the star exploded, but it was irrelevant to the drama about to take place in the core. The week-long fury of silicon- and sulfur-burning had left the star with a core of iron, together with other elements in the iron group: nickel, chromium, titanium, vanadium, cobalt and manganese. Vast neutrino losses continued unabated because of the high core temperature, but having reached iron, the core had no nuclear currency left to pay its energy debt. Iron lies at the bottom of the curve of binding energy: energy must be added to fuse it into heavier elements or to split it into lighter ones. Fusion could go no further, and temperature and pressure could no longer maintain the core's equilibrium. Gravity won the 11-million-year contest, and the core began to collapse.

As the core was compressed, it did get hotter but not hot enough to stop the collapse. Two instabilities (discussed by William A. Fowler of the California Institute of Technology and Fred Hoyle, then of Cambridge University, during their pioneering theoretical work on supernovas in the early 1960's) actually accelerated the collapse. In one process, photodisintegration, high-energy photons tore apart the iron nuclei into lighter components, mainly helium—in effect reversing the fusion reactions of the star's previous history. In the second process, electron capture, free electrons were squeezed into nuclei, where they combined with protons to form neutron-rich isotopes. Both processes consumed energy, sapping critical support from the core; electron capture also removed some of the free electrons that had been a major source of pressure.

In a few tenths of a second the iron core, 1.4 times the mass of the sun and half the size of the earth, collapsed into a ball of nuclear matter about 100 kilometers in radius. When the center of the incipient neutron star exceeded the density of an atomic nucleus—270 trillion grams per cubic centimeter—the inner 40 percent of the core rebounded as a unit. The outer core, still plunging inward at close to a quarter of the speed of light, smashed into the rebounding inner core and rebounded in turn. A shock wave was born. In about a hundredth of a second, it raced out through the infalling matter to the edge of the core (see Chapter 9, "How a Supernova Explodes," by Hans A. Bethe and Gerald Brown).

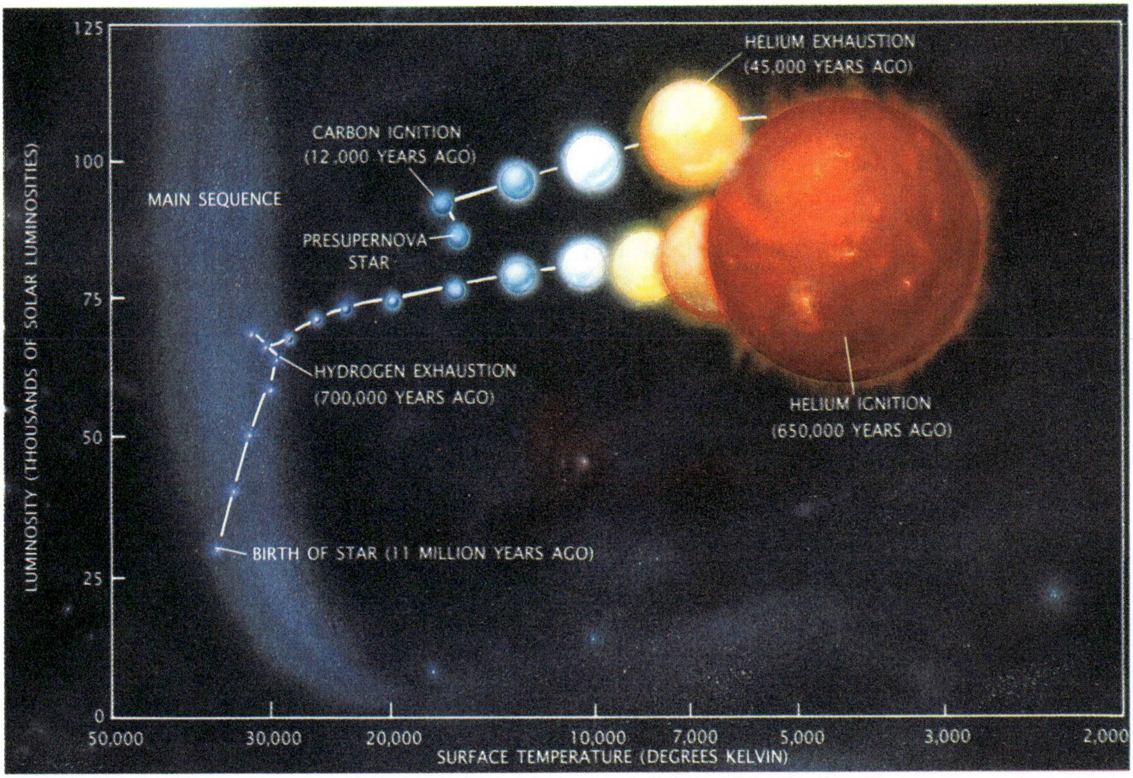

Figure 11.3 HISTORY of the progenitor star began some 11 million years ago. After about 10 million years the hydrogen was burned to helium, and the core contracted and became hotter. In response the star's envelope expanded and cooled, and the star moved to the right. As the core got hot and dense enough to burn helium, the star bloated into a red supergiant, with a cool envelope several times the size of the earth's orbit. After helium was exhausted, the envelope contracted and heated up again, and the star became a blue supergiant. In that form it burned successively heavier elements, ultimately making the iron core whose collapse triggered the supernova.

Workers modeling supernovas had hoped for many years that such a shock would continue outward through the many layers of the star, heating it and blowing it apart. Unfortunately, the most recent calculations for a star the size of Sk −69° 202, done by a number of theorists (including Sidney Bludman and Eric Myra of the University of Pennsylvania, Stephen Bruenn of the Florida Atlantic University, Edward A. Baron of the State University of New York at Stony Brook and Ron Mayle and James R. Wilson of the Lawrence Livermore National Laboratory), suggest that in SN 1987A the shock did not make it out of the core on its own.

The shock started out carrying enormous energy —about 10 times as much as was finally imparted to the exploding debris—but lost most of it beating outward against the infalling material. Photodisintegration and neutrino emission cooled the shock-heated material, sapping the shock's impetus. By the time the shock arrived at the edge of the iron core, the material behind it had no net outward velocity. The shock stalled and became an accretion shock, one through which material continuously flows inward. If this dismal state had persisted, the core would have swallowed the entire star. The result would have been a black hole, not a supernova.

Neutrino emission played a role in stalling the shock, and neutrino emission may also have helped to revive it. The core, having shrunk to a radius of 100 kilometers, had not reached nuclear density except at the center. It would become a true neutron star only when it had contracted to a radius of about 10 kilometers. Yet the protoneutron star was already very hot (Wilson and other modelers had predicted a temperature of about 100 billion degrees K.) because of the gravitational energy released in the collapse. To contract further, the neutron star had to lose heat.

It did so through vast neutrino losses. The neutrinos were produced, as before, by the annihilation of electron-positron pairs made by the energetic gamma rays that pervade material at such high temperatures. This time, however, the neutrinos did not stream promptly out of the material: the density of the collapsing core was so high that it impeded even neutrinos. They diffused out of the core gradually, in seconds rather than milliseconds, slowing the star's contraction.

Even so, the power radiating from the contracting neutron star was outrageous, exceeding that of the rest of the visible universe. The total energy emitted in the 10-second neutrino burst was 200 or 300 times the energy of the supernova's material explosion and 30,000 times the energy of its total light output. It is now widely (but by no means universally) believed that a small fraction of the neutrino energy was somehow harnessed to revive the stalled shock and power the explosion. Building on a basic idea put forward in the mid-1060's by Stirling Colgate of the Los Alamos National Laboratory, Mayle and Wilson did a set of calculations that show just such an effect. Only a few percent of the neutrinos, interacting with the material just behind the stalled shock for about a second, deposit enough energy to accelerate the shock outward (see Figure 11.4).

By heating and expanding the star and triggering a new flurry of nuclear reactions in its layered interior, the revived shock was responsible for the supernova's optical display. The effect was delayed by about two hours: the shock traveled at perhaps a fiftieth of the speed of light and had to traverse the entire star before any light leaked out. The neutrinos from the collapsing core easily outraced the shock. Passing through the rest of the star very close to the speed of light, they were the first signal to leave the supernova.

Some 160,000 years later, still hours ahead of the light front, the neutrinos swept over the earth—

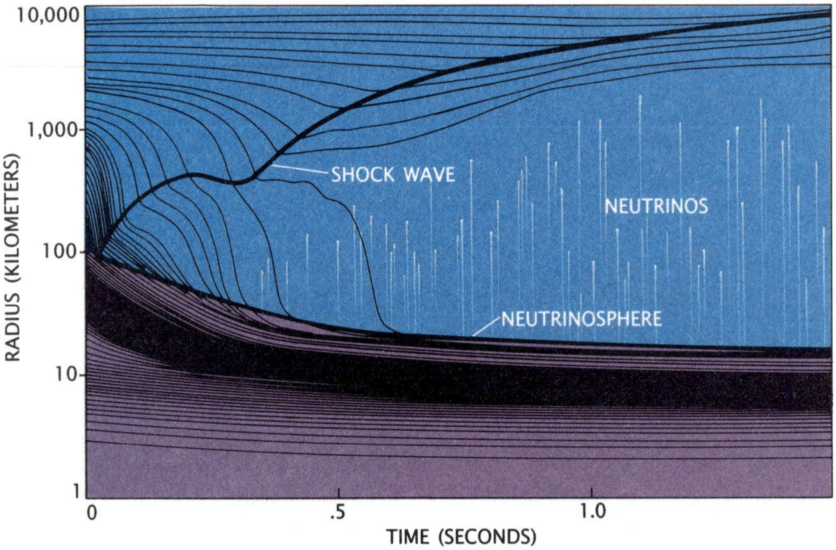

Figure 11.4 NEUTRINOS REVIVE THE SHOCK WAVE generated by core collapse. Each line on the graph traces the radial position of a shell of constant mass. As the time scale begins, the shock has lost energy and stalled within the infalling material of the outer core. The graph shows how the collapsed core (*purple*)—a protoneutron star—contracts further and emits a powerful flux of neutrinos, which escape from its surface (the "neutrinosphere") after diffusing through the nuclear matter. A trace of energy deposited by neutrinos heats and accelerates material behind the shock. The revived shock is sufficient to destroy the star. (Calculations by Ron Mayle and James R. Wilson.)

and were detected. Investigators searching for rare subatomic events such as the decay of the proton have built detectors deep in mines and under mountains, where they are shielded from interference by cosmic rays. Typically they consist of a swimming-pool-size tank of water flanked by arrays of photodetectors, poised to sense the faint flashes of light that would signal the decay of any one of the perhaps 10^{32} protons in the water. To date no proton has been seen to decay, but the detectors are also sensitive to another rare, energetic event, the capture of a neutrino by a proton.

On February 23 at 7:36 A.M. Universal Time, the Kamiokande II detector, in the Kamioka lead mine in Japan, and the IMB detector (named for the collaborating institutions, the University of California at Irvine, the University of Michigan at Ann Arbor, and the Brookhaven National Laboratory) in the Morton Thiokol salt mine near Cleveland, Ohio, simultaneously recorded a series of events that were later interpreted as neutrino captures. A detector of a different design, at Baksan in the Soviet Union, registered anomalous events at the same time. Approaching from out of the southern sky, the wave of neutrinos from the supernova had swept through the earth (the earth is far more transparent to these weakly interacting particles than a thin sheet of the clearest glass is to light). Emerging in the Northern Hemisphere, it had left the faintest signature of its passage in the detectors.

The theoretical significance of the neutrino detection was considerable. The Kamiokande and IMB detectors are most sensitive to a small component of the burst: electron antineutrinos. The same proportion of the burst energy is believed to have come from each of the other five neutrino flavors—electron neutrinos and mu- and tau-neutrinos and their two antiparticles. By extrapolating from the number and energy of the neutrinos that were detected, workers have calculated the total neutrino energy released by SN 1987A: 3×10^{53} ergs. It is just equal to the theoretical binding energy of a neutron star of 1.4 solar masses—the gravitational energy that should be released in its formation.

Thus, the fleeting detection of the neutrino burst shows that, as theory had predicted, a neutron star is formed in a type II supernova. More specifically, it is a sign that computer models of the formation and collapse of massive stars are on the right track: they had accurately predicted the mass of the imploding core. The average energy of the detected neutrinos confirms theoretical predictions for the temperature of a collapsing protoneutron star. Furthermore, the burst lasted several seconds; the neutrinos actually did have to diffuse out of the dense matter of the collapsed core.

Of even broader significance was the fact that the neutrinos arrived as a close-packed bunch a few hours ahead of the light burst after a journey of 160,000 years. The universe is widely believed to contain far more mass than can be seen, and neutrinos have been proposed as the carrier of this "missing mass." The fact that the neutrinos traveled so close to the speed of light sets strict limits on their mass: neutrinos with significant mass traveling at such speed would have been far more energetic than the detected particles. Furthermore, neutrinos of quite different energies arrived within seconds of one another; in contrast, the arrival times of particles with significant mass would have been spread out in order of decreasing energy.

Independent analyses of the timing by John Bahcall of the Institute for Advanced Study in Princeton, N.J., Adam Burrows of the University of Arizona and Tom Loredo and Don Lamb of the University of Chicago have yielded a firm upper limit on the electron antineutrino's mass: about 20 electron volts (.00004 times the mass of the electron). If the masses of mu- and tau-neutrinos could be limited to similar values, neutrinos could confidently be dismissed as a missing-mass candidate.

The neutrino burst bore tidings of the core collapse, but it had very little to say about how the shock generated by the collapse got out of the core. The revival of the shock by neutrino energy remains in the realm of theory. Nevertheless, it is clear that a strong shock did propagate through Sk $-69°$ 202 on February 23, 1987 (minus 160,000 years). To state the obvious, the star did explode.

Two hours after the neutrinos had been registered in the Kamiokande and IMB detectors (nobody knew it at the time, of course), Albert Jones, an amateur astronomer in New Zealand, happened to be observing the exact spot at which the supernova would appear. He did not see anything unusual. An hour later, in Australia, McNaught exposed the two plates that, when they were developed after Shelton's announcement of the discovery, showed the earliest recorded light from the supernova. Sometime between the two observations, perhaps even as Jones observed the spot, the shock erupted through the surface of the star, triggering a hard (short-wavelength) ultraviolet burst that quickly gave way to visible light.

The fact that it took only about two hours after the core collapse for the shock to arrive at the surface and ignite the optical display helped to dispel initial doubts about whether the blue star Sk $-69°$ 202 really was the star that exploded. The quick arrival of the light ruled out a red supergiant as the progenitor: it takes even a high-velocity shock the better part of a day to go through a red supergiant.

Further evidence about the size of the progenitor star came from the ultraviolet flash, even though only its aftermath was seen. In addition to being invisible, ultraviolet light is absorbed by the earth's atmosphere. The telescope on board the *International Ultraviolet Explorer* satellite could have detected this earliest light but was not aimed in the right direction at the time. Within 14 hours, however, the observing team, headed by Robert P. Kirshner of the Harvard-Smithsonian Center for Astrophysics and George Sonneborn of the National Aeronautics and Space Administration's Goddard Space Flight Center, had reoriented the satellite. By that time the initial burst was fading, but the supernova was still clearly visible at ultraviolet wavelengths.

Moreover, the workers got an indirect look at the ultraviolet flash months later, when the *IUE* detected emissions from a shell of gas surrounding the supernova at a distance of about a light-year. The gas, presumably material ejected from the presupernova star in a stellar wind during its red-supergiant stage 40,000 years before, was flash-ionized when the intense ultraviolet burst reached it. Based on this secondary radiation, Claus Fransson of the University of Stockholm concluded that the first light from the supernova came from material at a temperature of about half a million degrees K. (Perhaps 10 years from now, according to a model developed by Roger A. Chevalier of the University of Virginia, the shell will radiate again, this time in the radio and X-ray bands, when the supernova ejecta finally collide with it.)

Such high temperatures, and the very hard ultraviolet radiation they produce, are expected when a powerful shock wave breaks through the surface of a relatively small progenitor star. With less surface area in which to deposit its energy, the shock generates a correspondingly higher temperature, and it also accelerates the material to higher velocities. Doppler-shifted lines in the early ultraviolet and optical spectra indicated that the material had been ejected from the star at roughly one tenth the speed of light.

This expansion cooled the outermost layers of the young supernova, and the dominant emissions quickly shifted from the ultraviolet to the cooler, visible wavelengths recorded in the earliest photographs. The bolometric luminosity (the combined radiation at all wavelengths from the infrared through the ultraviolet) declined steeply during these first hours, but because the visible portion of the emissions was strengthening, the supernova was brightening into an impressive display in the night sky.

During the first day or so, little radiation could escape from deep inside the supernova: free electrons in the ionized gas of its envelope scattered light from deeper layers. When the outermost material had cooled to about 5,500 degrees K., however, the hydrogen nuclei recombined with the free electrons. As the supernova continued to expand and cool, a surface defined by the hydrogen recombination temperature moved into the envelope. At this surface energy previously deposited by the shock was released—mostly at visible wavelengths—and streamed freely into space. For weeks to come, as Arnett and Sydney W. Falk of the University of Texas at Austin had predicted some 15 years ago, radiation at the hydrogen recombination temperature dominated the supernova's emissions.

At the same time another effect of the small progenitor star became apparent. As an optical display the supernova was at first unexpectedly faint—about a tenth as bright as other type II supernovas at a similar stage. To cool to the hydrogen recombination temperature, any supernova has to expand. The shock had deposited about the same amount of energy in the relatively small envelope of this progenitor star as it would have left in a red supergiant's extended envelope, heating the small envelope to a correspondingly higher temperature. As a result SN 1987A had to expand by a much larger factor before it could release its light, and the process consumed energy that would otherwise have come out as radiation.

After about a month, it is calculated, all the energy deposited by the shock had either escaped as radiation or gone into accelerating the ejecta. Yet the supernova was still brightening at visible wavelengths. By this time, in April, another source of energy was providing most of the light: the decay of radioactive isotopes produced in the explosion. Most theorists had expected such materials to be made in a type II supernova, but they watched eagerly to see how much had been generated in SN 1987A and what role the isotopes would play.

The shock wave's passage through deep layers of the progenitor star during the first minutes of the event had triggered new nuclear reactions (see Figure 11.5). In particular, part of the silicon shell was turned into iron-group elements, chiefly the radioactive isotope nickel-56. A month later the highly unstable nickel-56 had already decayed (its half-life is 6.1 days), heating and expanding the deep layers of the supernova. But its decay product, cobalt-56, is also radioactive, and because it has a half-life of 77.1 days, it was still abundant. It decays into an excited iron-56 nucleus, which releases gamma rays at specific energies as it relaxes to the ground state. These gamma rays now powered the display.

At first the gamma rays themselves did not escape: because of their high energy they scattered repeatedly from electrons in the expanding gas, turning into X rays of progressively lower energy. At a sufficiently low energy the X rays were absorbed, heating the material and so contributing to the optical display. As the supernova continued to

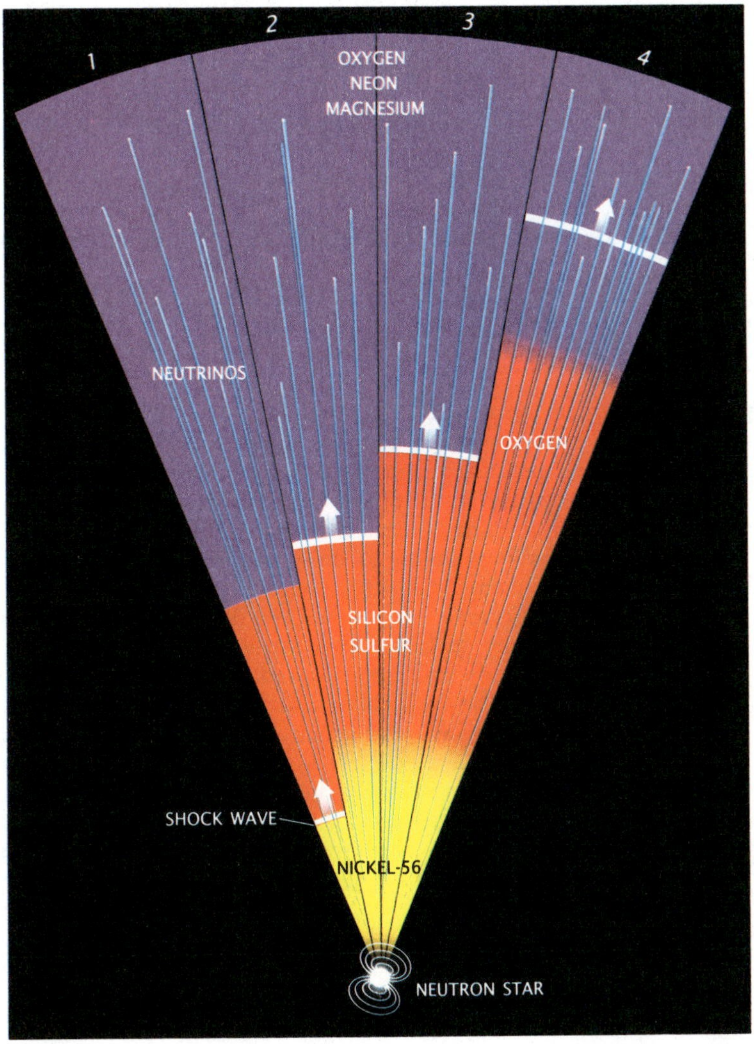

Figure 11.5 EXPLOSIVE NUCLEOSYNTHESIS takes place as the shock wave rips out through the progenitor star's layered interior. Shock-heated to more than five billion degrees K., part of the silicon and sulfur fuses to form radioactive nickel-56 (*stage 1*); some of the oxygen at the bottom of the next shell burns to silicon and sulfur (*stage 2*), and neon and magnesium in the inner part of the shell burn to oxygen (*stage 3*). The shock propagates through the remaining material without triggering further transmutations (*stage 4*). Neutrinos from the hot, contracting neutron star outrace the shock wave.

thin, increasing amounts of the decay energy escaped in this way. On May 20, 80 days after the explosion, the brightness peaked.

By early July the light curve was declining at precisely the rate expected on the basis of cobalt-56's half-life. From the brightness of the supernova on a given day and the time since the explosion, it was straightforward to calculate how much nickel-56 had been formed in the first place. The answer, .08 solar masses, is within a factor of two of what we and others had predicted for type II supernovas.

For weeks after peak brightness the radioactive material still could not be seen directly. By August, however, the expanding debris had thinned enough to allow some radiation from the decay to escape with little or no scattering. First the Japanese satellite *Ginga* and, shortly thereafter, instruments on the Soviet space station *Mir* detected X rays at the energies that Philip A. Pinto of the University of California, Santa Cruz, Rashid A. Sunyaev and S. A. Grebenev of the Soviet Space Research Institute, and others had predicted would be seen when the gamma rays from cobalt-56 decay were scattered. Once the X rays had been seen, the gamma rays themselves could not be far behind, and in December their discovery was announced based on data from the *Solar Maximum Mission* satellite (see Figure 11.6). Confirmation came quickly from balloon-borne detectors flown in Australia and in Antarctica.

Donald D. Clayton of Rice University and his co-workers had predicted some 20 years ago that a supernova would produce gamma rays of the observed energies, but the timing of the observations was a surprise. Theorists had expected that in a type II supernova the layers of the exploded star would expand in radial symmetry, in which case the X rays would have been obscured until perhaps 100 days after they were actually observed. Their early appearance meant, instead, that the core had been mixed: material from the inner layers had been blown into the overlying layer of helium or even into the hydrogen envelope. Indeed, Doppler broadening of the gamma-ray lines showed that some cobalt was moving as fast as 3,000 kilometers per second—fast enough to have overtaken the slower-moving material at the base of the hydrogen envelope.

At about the time the cobalt appeared, emissions from deeper in the supernova revealed other heavy elements. Gamma rays and X rays from the core of the supernova were still being scattered, and visible and ultraviolet emissions were blocked by a thicket of atomic absorption lines. The infrared, it turned out, offered the earliest look at the heavy elements the supernova was dispersing into space (see Figure 11.7).

Most infrared radiation is absorbed by the earth's atmosphere, but the wavelengths that do reach the ground were studied, beginning soon after the supernova exploded, by the Anglo-Australian Telescope at Coonabarabran and the Mount Stromlo and Siding Spring Observatories in Woden (both in Australia) and by the Cerro Tololo Inter-American Observatory in Chile. NASA's Kuiper Airborne Infrared Telescope, flown at 39,000 feet on a jet transport, gained more complete coverage on flights starting in the fall of 1987. Beginning around November, spectra from the Kuiper and from Australia together revealed an entire zoo of elements in the supernova core—not just iron, nickel and cobalt but also argon, carbon, oxygen, neon, sodium, magnesium, silicon, sulfur, chlorine, potassium, calcium and possibly aluminum. Their intense infrared lines signaled large quantities than could have been present in the star at its birth. The elements—the components, perhaps, of some future solar system—were made in the core of the star or in the explosion itself.

In early 1989, two years after the explosion, the supernova's luminosity was declining steadily, in keeping with the exponential decay of radioactive cobalt-56 (save that some of the X rays and gamma rays could now escape directly and hence did not contribute to the light curve). The lack of evidence for any energy source other than radioactive decay was starting to puzzle some theorists. The neutrino burst had announced the birth of a neutron star. Yet a neutron star usually emits a great deal of radiation, either by heating any material falling into it or by acting as a pulsar: a spinning neutron star with a strong magnetic field that generates a rotating beacon of radiation.

Where was the neutron star in SN 1987A? Had it formed initially but then vanished by turning into a black hole? The neutrino burst would have been cut short if a black hole had formed during the first few seconds of the event, and in any case the mass of the collapsing iron core alone fell short of the threshold—about two solar masses—for forming a black hole. If enough additional mass had later fallen onto the neutron star to drive it over the limit,

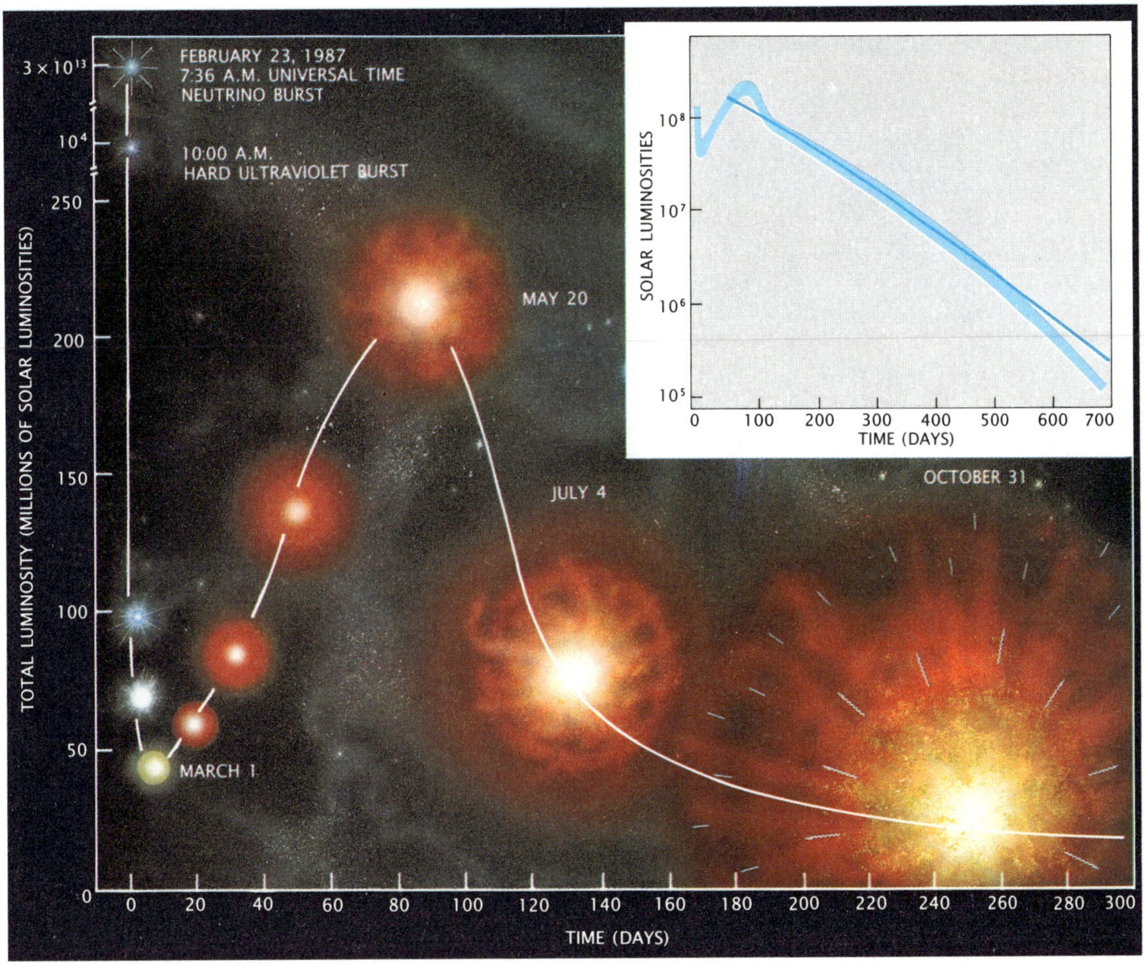

Figure 11.6 EXPLOSION of the supernova began with a powerful burst of neutrinos, followed some two hours later by a flash of hard ultraviolet light as the shock wave broke through the surface of the star, heating it to half a million degrees K. The supernova brightened slowly until May 20, by which time the shock energy had been spent and the display was powered entirely by radioactivity. The subsequent decline in brightness, plotted on a logarithmic scale (*upper right*), matched the decline calculated for the decay energy of .08 solar masses of cobalt-56 (*dark curve*). Months after the explosion, as the supernova thinned into a clumpy nebula many times the size of the solar system, X rays and gamma rays (*blue wavy lines*) from the decay of the cobalt began to escape directly.

all the radioactive nickel would have been lost and the supernova would have been much fainter. As the supernova neared its second anniversary, most astronomers were still betting on a neutron star, although the exponential decline of the light curve ruled out a very bright pulsar such as the one in the Crab Nebula, the remnant of a brilliant supernova in 1054.

During the night of January 18, 1989, Universal Time, the supernova answered one puzzle with several more. At Cerro Tololo a group headed by Carl Pennypacker of the Lawrence Berkeley Laboratory and John Middleditch of Los Alamos detected optical pulsations from the supernova. The pulsations, which amounted to about .1 percent of the total light, came nearly 2,000 times a second, suggesting a rotation rate three times faster than the fastest pulsar ever seen. Spinning that fast, only the den-

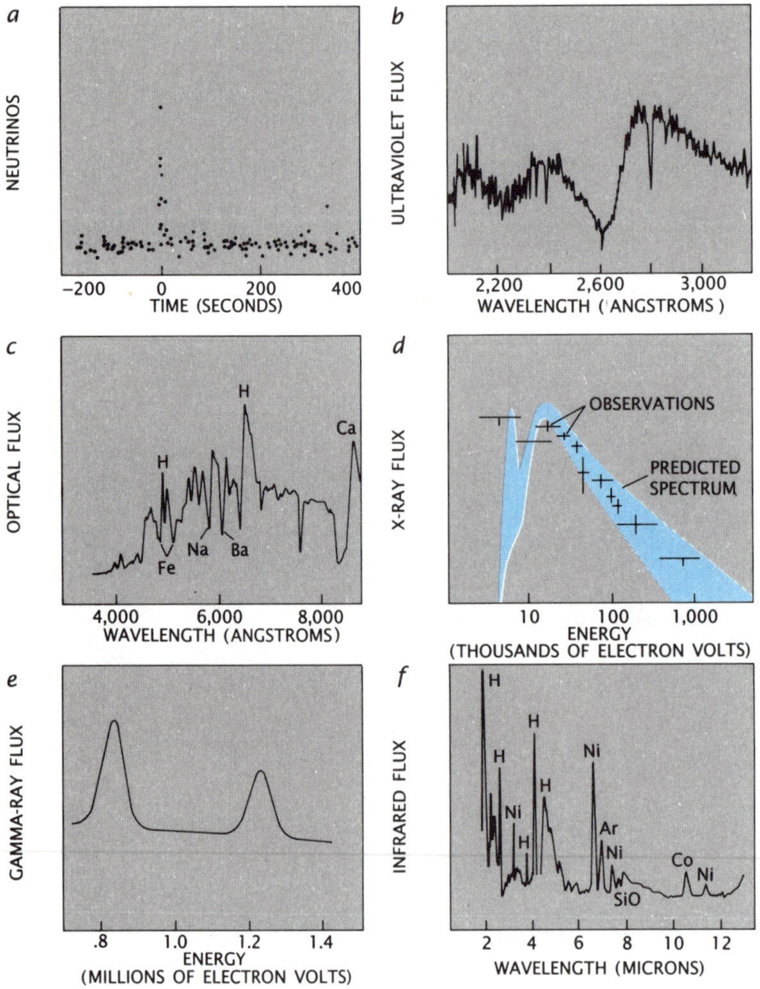

Figure 11.7 EMISSIONS from SN 1987A began with a brief burst of neutrinos, shown in a record from the Kamiokande detector (*a*). Hours after the shock wave burst from the star, the *International Ultraviolet Explorer* satellite recorded an ultraviolet spectrum testifying to the very high temperature of the shock-heated surface (*b*). A spectrum at visible wavelengths, made 50 days after the explosion, shows strong spectral lines of hydrogen, characteristic of the expanding, cooling envelope (*c*). After about six months instruments on the satellite *Ginga* and the space station *Mir* detected X rays from the decay of radioactive cobalt (*d*); the detection of gamma rays from the same decay by the *Solar Maximum Mission* satellite was reported a few months later (*e*). Infrared emission lines captured by the Kuiper Airborne Infrared Telescope reveal a variety of newly made elements deep in the expanding ejecta (*f*). (Ultraviolet spectrum provided by Robert Kirshner, gamma-ray data by Mark D. Leising and the infrared spectrum by Fred C. Witteborn.)

sest, most massive neutron star allowed by theory could avoid flying apart.

What is more, the signal of the pulsar showed a regular variation in frequency, as if an orbiting companion several times as massive as Jupiter were tugging the pulsar back and forth every seven hours, Doppler-shifting its signal. Because the radius of the companion's calculated orbit, about a million kilometers, would have placed it inside the presupernova star, the companion could only have been created after the explosion. Speculation is rife: if the companion is real, could it be a piece of neutron star that was somehow ejected, some other fragment that fell back and was captured, or something even more exotic?

What is really needed is another look at the pulsar. Yet several months of observations of equal and greater sensitivity have failed to recover it. Again, one can speculate. Clouds deep in the supernova may be obscuring the pulsar, for example, or it may have been extinguished: matter falling onto the neutron star may have short-circuited the electric field (generated by the rotating magnetic field) that powers the beam. No one knows.

The fate of the neutron star joins other mysteries that have accompanied SN 1987A. We have emphasized the success of theory and the beautiful complementarity among the observations. Yet there had been anomalies even before the putative pulsar.

Four hours *before* the neutrino detection at Kamiokande and IMB, for example, a detector in a tunnel under Mount Blanc had registered a separate neutrino burst. Gravity-wave detectors (sensitive to massive releases of gravitational energy) in Rome and in Maryland are said to have recorded signals coincident with those early neutrinos. What could account for a stupendous burst of energy four hours before the core collapse? Again, no one knows. Several months after the explosion came another mystery: a second light source, roughly one tenth as bright as the supernova and resolvable from the main explosion only by an indirect technique known as speckle interferometry. The mysterious second source had disappeared by June, 1987, and was not seen again.

Doubts about such observations, and controversy about their interpretation, bring home an important point about the supernova. In much of science a result is accepted only if it is reproducible. Yet in the case of supernova 1987A we deal with an event that may not be repeated nearby for centuries. When our ability to interpret the observations breaks down, the best we can do is to record and archive the findings carefully, so that future scientists, with greater insight, may come to understand them.

Even so, the last two and a half years have yielded breathtaking advances in the understanding of type II supernovas. For us and hundreds of others, theorists and observers at all wavelengths collaborating to document and explain one of the heavens' grandest events, it has been a time of matchless exhilaration, scientific co-operation and intellectual reward—the event of a lifetime.

The Oldest Pulsars in the Universe

12

These unusual pulsars are dense, compact stars spinning at the rate of several hundred revolutions per second. Why do they spin so fast? They are thought to have been resurrected from an early death.

• • •

Jacob Shaham
February, 1987

One sunny morning in the middle of November, 1982, David J. Helfand, an astronomer colleague at Columbia University, came into my office to tell me a remarkable story. He had just returned from a business trip to the Arecibo radio-telescope observatory in Puerto Rico. While he was there he learned that Donald C. Backer of the University of California at Berkeley and his collaborators had discovered a 1.558-millisecond radio pulsar in the constellation Vulpecula. That would make the radio pulsar, whose official name is 1937 + 214, the fastest one known.

Backer's discovery was a startling piece of news indeed. Radio pulsars are starlike celestial objects that generate many kinds of particulate and electromagnetic radiation. Among them are radio waves. When a radio telescope is directed at such a source, the radio waves are registered as periodic "beeps." The fastest radio pulsar known prior to the discovery of 1937 + 214 was the Crab pulsar: a 30-millisecond-period beeper buried inside a large, fuzzy structure resembling a cotton ball and called the Crab Nebula. Now there appeared to be something beeping 20 times faster.

Since the discovery of the Vulpecula pulsar two more superfast pulsars have been reported. In 1983 a 6.13-millisecond pulsar (called 1953 + 29) was announced, and in 1986 a 5.362-millisecond pulsar (called 1855 + 09) was publicized. A candidate for a fourth has been mentioned. As more evidence becomes available, it seems increasingly likely that the superfast pulsars can be explained only as a part of a new class of pulsars.

Although many of the details of the class remain obscured, some general facts are emerging. Perhaps most interesting of all is the great age these new celestial objects are thought to have. Ordinary pulsars are relatively young, typically less than a million years old; the Crab pulsar, which is the youngest one known is a mere infant of 932 years. The superfast pulsars, in comparison, are thought to be ancient. They are probably the result of evolutionary processes that could go back as much as a billion years, or one-twentieth of the age of the universe, and they are likely to live for several billion years more.

Ordinary radio pulsars are fascinating in themselves. They are thought to be rotating neutron stars: huge, spinning "nuclei" that contain some 10^{57} protons and neutrons. (The nucleus of a hydrogen atom, in contrast, consists of a single proton; an iron nucleus contains a total of 56 protons and neutrons.) The neutron star gets its name from the fact that it has about 20 times more neutrons than protons. The large clump of nuclear matter, which has a mass about equal to that of the sun, is compressed into a sphere with a radius on the order of 10 kilometers. Consequently the density of the star is enormous, slightly greater than the density of ordinary nuclear matter, which is itself some 10 trillion times denser than a lead brick (see Figure 12.1).

Currents of protons and electrons moving within the star generate a magnetic field. As the star rotates, a radio beacon, ignited by the combined effect of the magnetic field and the rotation, emanates from it and sweeps periodically through the surrounding space, rather like a lighthouse beam. Once per revolution the beacon cuts past the earth, giving rise to the beeping detected by radio telescopes. The Crab pulsar, for instance, is a star as massive as the sun squeezed into a shell the size of an average city and spinning at the high rate of 33 times per second. (Every 30 milliseconds the Crab pulsar makes one complete revolution.)

One can amuse oneself for many hours thinking about the properties of a neutron star. One can easily calculate, for example, the force of gravity at the surface of a neutron star. It turns out to be some 100 billion times greater than the force of gravity at the surface of the earth. Indeed, the gravitational field is so strong that any object drawn toward a neutron star would fragment even before impact: objects cannot fall on a neutron star, they can only "rain" on it.

What processes could give rise to the birth of such an extraordinary star? Astrophysicists believe a neutron star is formed when an ordinary star, which

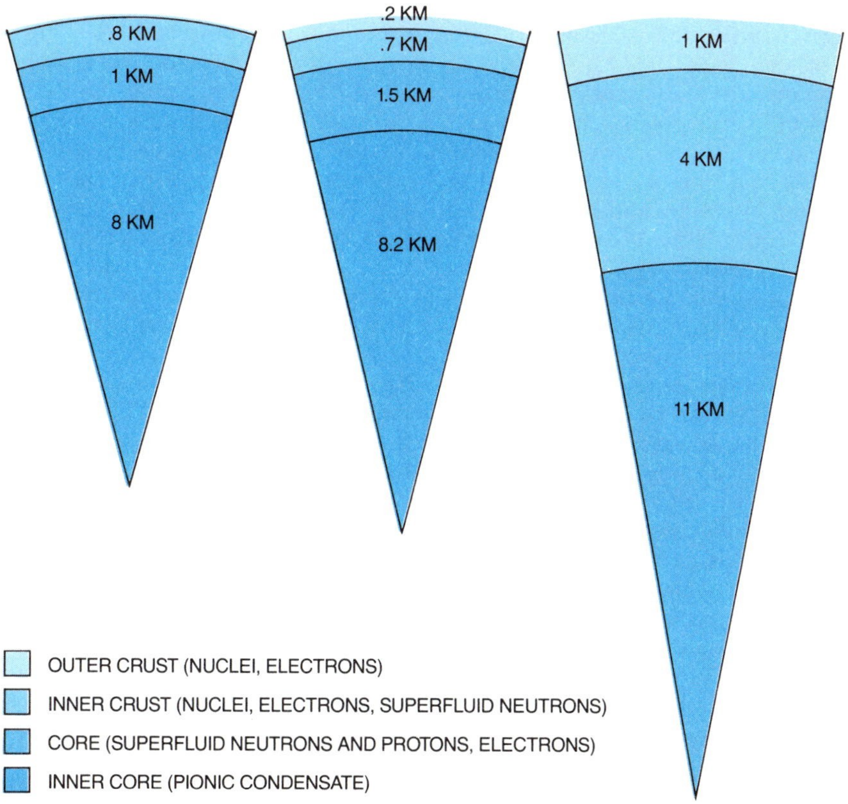

Figure 12.1 NEUTRON STAR is shown in cross section for each of three theoretical models. A neutron star consists of neutrons and protons packed so tight that its density exceeds even that of ordinary nuclear matter, which is itself some 10 trillion times denser than a lead brick. The star depicted here has a mass 1.4 times as great as the sun's, yet its radius is only somewhere between 10 and 16 kilometers, which is the size of an average city.

has a mass several times greater than that of the sun, collapses violently. The event is called a supernova. The inner parts of the ordinary star implode to form the neutron star while the outer parts explode to form a surrounding nebula, or cloud of gas and dust. Some of the energy from the explosion is released as visible light. The display can sometimes be quite spectacular. In A.D. 1054, for instance, when the supernova that gave birth to the Crab pulsar and Crab Nebula occurred, the light display was visible during the daytime for nearly three weeks and visible during the nighttime for more than a year. Energy continues to be released even after a pulsar is born: the spinning star ejects high-energy particles that are absorbed by the surrounding nebula, making it glow. As the pulsar ejects the particles it loses rotational energy and slows down. A general rule of thumb follows that the older the pulsar, the slower its rate of spin.

One might therefore expect that the recently discovered superfast pulsars must be very young—certainly younger than the 932 years of the Crab pulsar. In late 1982 and early 1983 M. Ali Alpar of the Research Institute for Basic Sciences in Turkey, Andrew F. Cheng of Johns Hopkins University and Malvin A. Ruderman of Columbia University joined me in analyzing the data from the superfast pulsar 1937 + 214 collected by Backer and his colleagues (S. R. Kulkarni and Carl E. Heiles of Berkeley, M. M. Davis of the Arecibo Observatory and W. M. Goss of the Kapteyn Laboratory in the Netherlands). We calculated, as others have, that the pulsar was rotating so fast that it was a factor of only two or so away from actually flying apart owing to centrifugal forces. It was clear that the star could not have been rotating much faster when it was born and therefore could not have slowed down by much.

A statistical consideration, however, would ultimately lead to the conclusion that even though the superfast pulsar has a high rate of spin, it is probably an object of great antiquity. In other words, it appears to have been formed during a long evolutionary process, not during the relatively quick explosion that is characteristic of a supernova.

The statistical consideration pertains to how fast a radio pulsar slows down. Simply stated, the shorter the period of a pulsar is, the more it slows down; the longer the period is, the less it slows down. As a consequence there should be more slow pulsars than fast ones. The existence of the 1.5-millisecond pulsar 1937 + 214, unless it is some freak accident, implies that if pulsars are born at a fairly constant rate over time, there should now be many pulsars that have slow periods of between 10 and 30 milliseconds. Yet in spite of a fairly extensive search of the heavens, no such pulsars have been found. Moreover, even though 1937 + 214 has such a short period, it is barely slowing down. Why has it continued to spin so fast?

To account for the discrepancy between theory and observation, we made use of the fact that the rate at which a pulsar slows down also depends on the strength of the magnetic field at the surface of the star. The greater the magnetic field is, the more energy the pulsar radiates and the faster it will slow down. A typical neutron star has a magnetic field of about a trillion gauss, which is roughly 100 million times greater than the field of a heavy-duty magnet. We showed that if a 1.5-millisecond pulsar had a magnetic field 1,000 times weaker (about a billion gauss), it could conceivably require a time comparable to the age of the universe to slow down to a period of 10 milliseconds (see Figure 12.2). This could explain why no pulsars have been seen in the 10-to-30-millisecond range: they simply would not yet have had time to slow down.

A few simple calculations also show that if the magnetic field of a pulsar is relatively weak, the star may even "turn off," or stop radiating, before it can slow down greatly. An ordinary pulsar—one that has a magnetic field of a trillion gauss—turns off when the spin rate is lower than about one revolution per second. A pulsar that has a magnetic field of 100 million gauss, in contrast, would turn off if its spin rate were lower than one revolution per 10 milliseconds (see Figure 12.3). In other words, even if a pulsar with a weak magnetic field could have somehow slowed down to 10 milliseconds, it would not be visible anyway.

All these considerations amount to a remarkable result: the mere existence of pulsar 1937 + 214 means there must be a new class of radio pulsars—very distinct from mainstream radio pulsars—that have weak surface magnetic fields.

The realization that there is a new class of radio pulsars with weak magnetic fields took a great load off our hearts. Observations of the radiation coming from 1937 + 214 and its vicinity indicate that the pulsar gives off comparatively little energy. Yet if an ordinary, high-magnetic-field pulsar were spinning as fast as 1937 + 214, its "slowing-down time" (the time required for its period to double) would be a few years and the star would radiate enormous

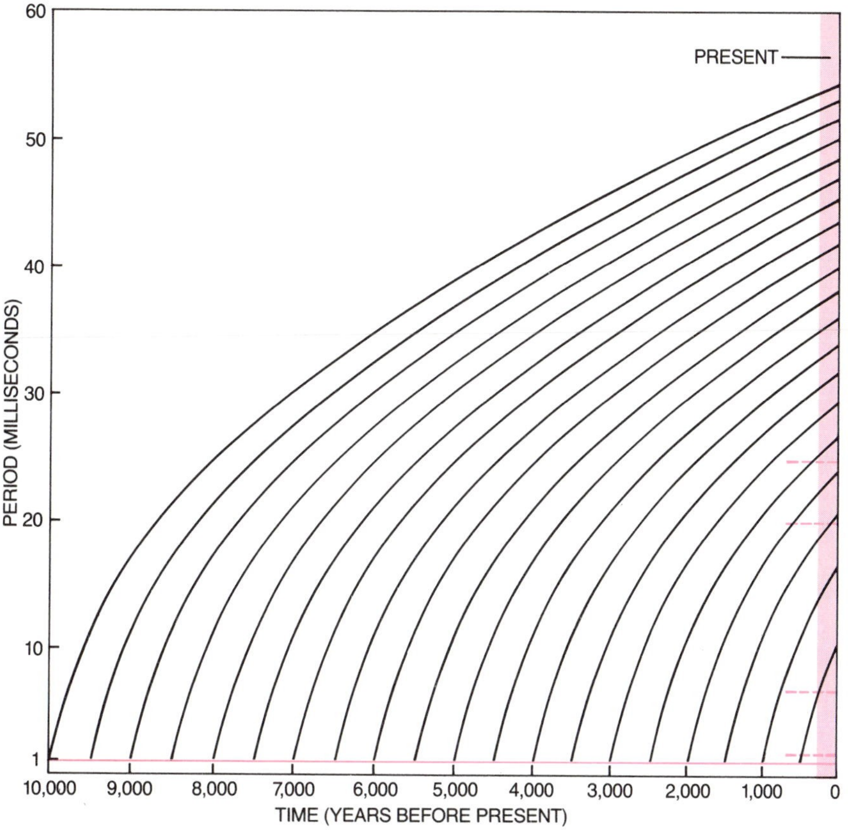

Figure 12.2 STATISTICS show that the millisecond pulsar 1937 + 214 must have a magnetic field that is much weaker than the field of an ordinary pulsar. The graph, which was prepared with the underlying assumption that the millisecond pulsar has a magnetic field of about a trillion gauss, traces the evolution of a family of millisecond pulsars (*black curves*). As a pulsar ages, its period becomes longer: it slows down. As a consequence the existence of the millisecond pulsar (*lower set of broken lines*) implies that there should be several pulsars with periods of 20 milliseconds (*upper set of broken lines*). No such pulsars have been found, however. The discrepancy can be explained by positing that the millisecond pulsars have weak magnetic fields, on the order of a billion gauss.

amounts of energy: it would be several million times more luminous than the Crab pulsar. We calculated that as a weak-field pulsar, 1937 + 214 has a slowing-down time of nearly a billion years and its energy output is less by a factor of some 100 million, which is consistent with observation (see Figure 12.4). Thus our theory successfully explains what would otherwise be an enormous discrepancy.

We were then faced with a new twist in our analysis. How could a pulsar with a weak magnetic field be created? It so happened that approximately two weeks before we heard about the fast spin of 1937 + 214, new work by Roger D. Blandford of the California Institute of Technology, James H. Applegate of Columbia and Lars Hernquist of Berkeley came to our attention. They showed that if a pulsar is formed at a high temperature—above about 100 million degrees Kelvin, which is actually below the characteristic temperature of a supernova—the pulsar will automatically acquire a high magnetic field. The implication of their conclusion was clear: pulsar 1937 + 214 must have been formed "cold" (at temperatures below 100 million degrees), which indicates that it could not have been created by a supernova. Why? Because a "hot" formation implies a large magnetic field and a rapid rate of slowdown, which means that a 1.5-millisecond pulsar with a small magnetic field could not have been formed.

The only known way that a cold neutron star could spin fast is if it accreted slowly rotating material. The basic phenomenon is familiar to anyone who has watched an ice skater achieve fast spins. First the skater spins slowly with outstretched arms, and then he or she draws the arms in, causing the spin rate to go up. For the same sort of process to occur with a cold neutron star the rate of accretion must be low or the star would overheat and the magnetic field would jump too high. As a consequence pulsar 1937 + 214 must have been in the making for a long time, certainly for at least 10 million years and probably on the order of a billion years.

The slow accretion could have proceeded in one of two ways. According to one scenario, a so-called white-dwarf star (which has the mass of a neutron star but a radius 1,000 times greater) accreted matter from a companion star orbiting it (see Figure 12.5). As the matter accumulated, the white dwarf collapsed into a neutron star. If the collapse were slow enough, the neutron star could have gradually "spun up" and become a millisecond pulsar. By accreting additional matter from its companion, the spin of the neutron star would have increased until it reached an equilibrium period.

Alternatively, a neutron star could first have been formed in the usual way (by a supernova) and behaved like a regular pulsar. After some 10 million years most of its crustal electric currents would have decayed, leaving a small magnetic field. If the "dead" pulsar were tightly orbited by a companion star, it could have slowly accreted the necessary

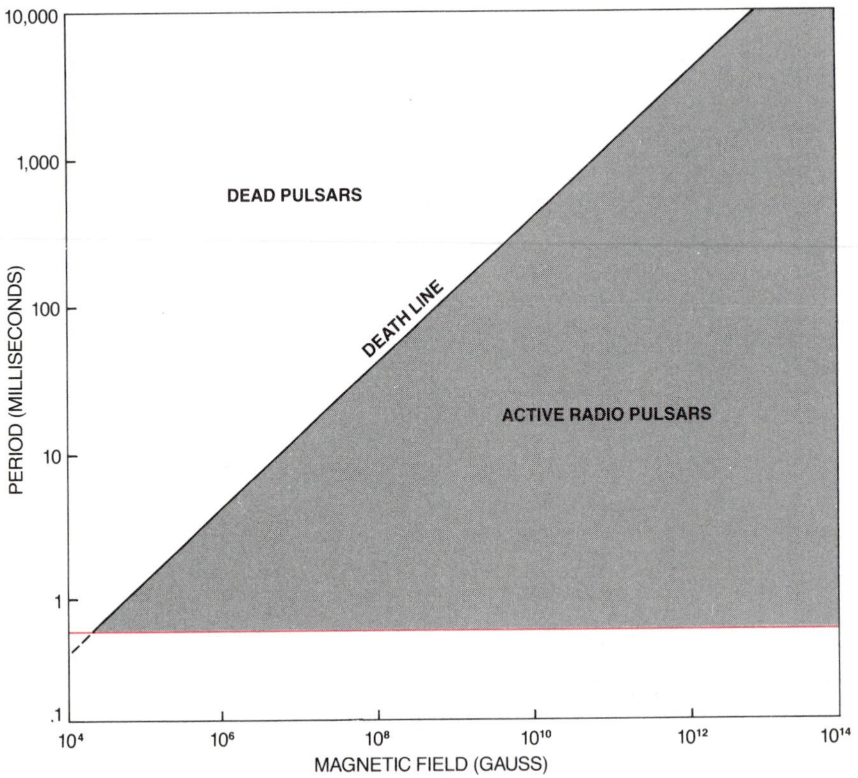

Figure 12.3 RADIO PULSARS turn off when their surface magnetic field is too weak or when their period of rotation is too long. The diagram shows the boundary between active and "dead" radio pulsars for various fields and periods. Note that because a neutron star will fly apart if its period is shorter than one millisecond, the surface field of an active pulsar must be at least 100,000 gauss, which is roughly 10 times the field of a powerful magnet.

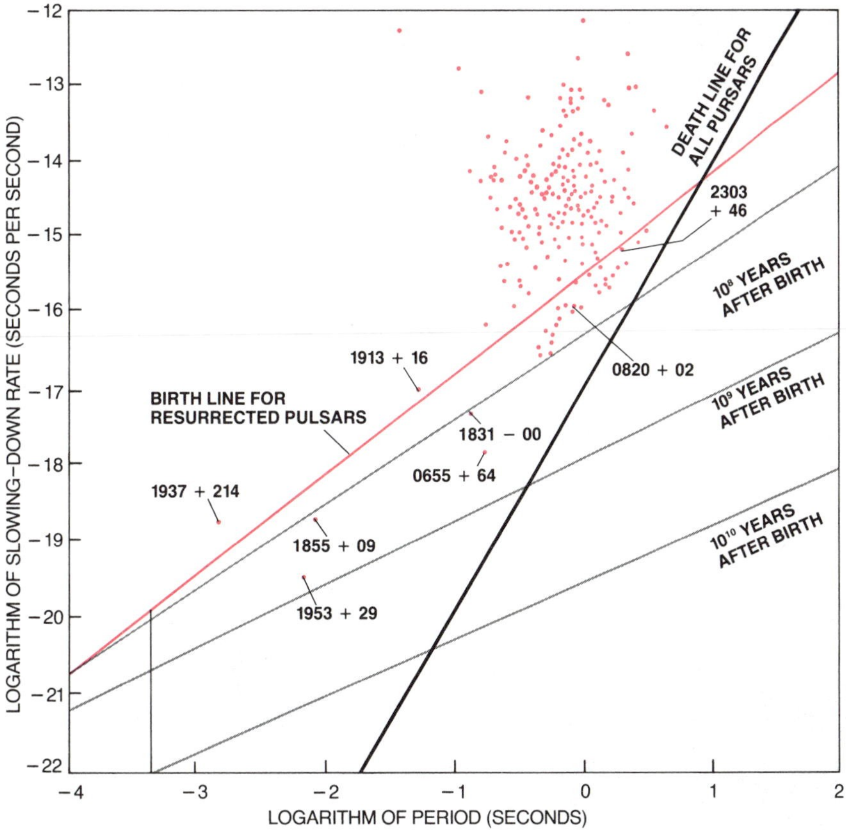

Figure 12.4 OBSERVED SLOWDOWN RATES of various pulsars are plotted against their observed periods. Most pulsars reside in the region of the diagram at the top right. As the pulsars age, they drift down and to the right until they hit the "death line" (see Figure 12.3). Six pulsars have been found that lie outside the main group. They (and two others that happen to lie within the main group) belong to a new class of pulsars that were spun up by a slow accretion process. All except 1937 + 214 have binary companions. The colored and gray diagonal lines designate various stages in the life of pulsars that were spun up at an accretion rate of 10^{-9} solar mass per year. The vertical line indicates the shortest period a neutron star can have before it flies apart.

material to acquire a fast spin and be resurrected, or recycled. Here and in the preceding scenario the companion would have to have been immolated in the process, since no star is found in the vicinity of 1937 + 214.

Underlying both scenarios, it is held, is a common phenomenon that guarantees the accretion will not proceed too quickly. As incoming electrons, atoms and molecules hit the surface of a star, some of their energy is transformed into light and other electromagnetic radiation. The radiation travels outward, striking other incoming electrons, atoms and molecules. Because radiation carries momentum, it exerts a force on the incoming matter. The radiation force is directed outward and therefore opposes the inward gravitational pull of the star.

If the accretion rate becomes high enough (6×10^{17} grams per second, or 10^{-8} solar mass per year), the radiation force is so strong that it balances the gravitational pull and the accretion rate cannot go higher. The accretion rate at which the two forces balance is called the Eddington limit, after the English astronomer Sir Arthur Eddington. In simple terms, the Eddington limit means that any accretion process onto a neutron star has a built-in thermostat that ensures the rate of accretion is never too high. If enough matter is available for accretion, the rate will also never be too low. The thermostat would

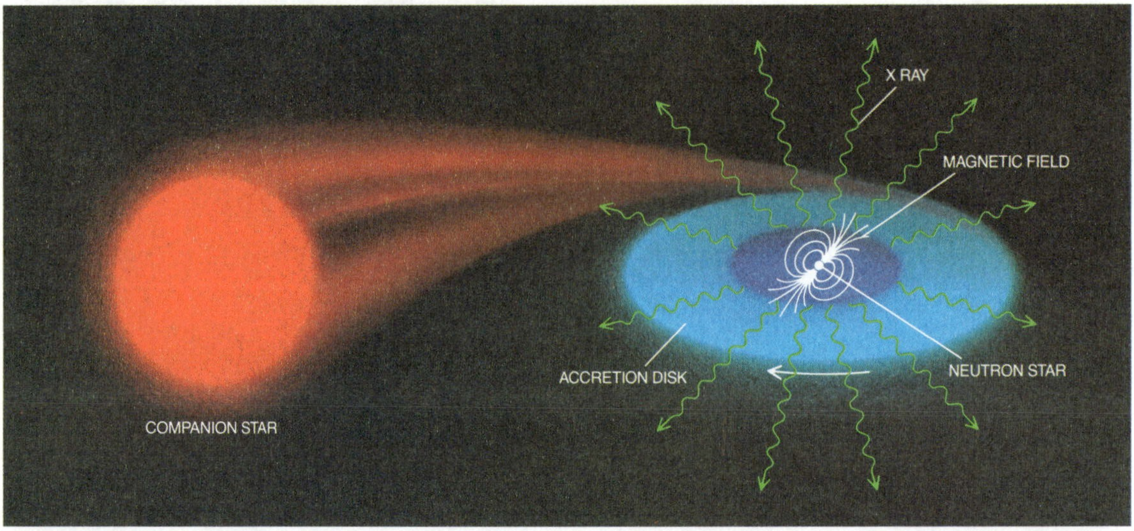
X RAY
MAGNETIC FIELD
ACCRETION DISK
NEUTRON STAR
COMPANION STAR

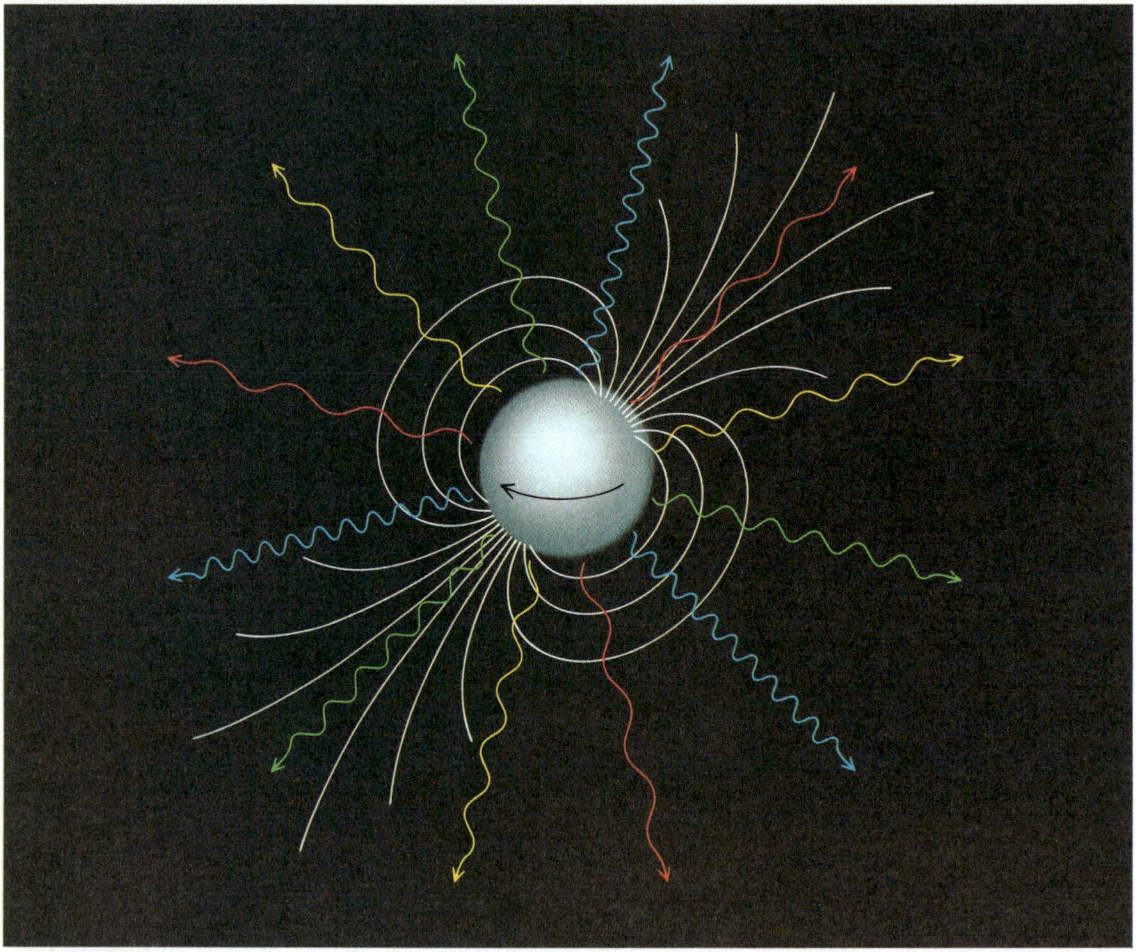

seem custom-tailored for the formation of millisecond pulsars, because limiting the accretion rate means limiting the stellar temperature to 100 million degrees or less.

Such an accretion-regulated process is already known to play an important part in the formation of "compact X-ray sources": small, rotating neutron stars that emit X rays. One kind of compact X-ray source is called a low-mass X-ray binary, in which the neutron star accretes mass from a light companion star. The accretion can last for 100 million years or more.

The fact that low-mass X-ray binaries have such long accretion epochs raised an interesting question: Could they be the progenitors of millisecond pulsars? We decided they simply had to be. Both the collapsing-white-dwarf scenario and the dead-pulsar scenario require a long-lived accreting binary star, and the only known candidate is the low-mass X-ray binary.

Statistical considerations seem to support the idea that low-mass X-ray binaries are the ancestors of millisecond pulsars. To give a flavor of the argument, I shall first discuss a simple example. Assume that all human beings live for 80 years. I shall define children as humans of age 10 or less; all the rest are adults. On the average, what should the ratio of adults to children be? Clearly there should be seven adults for every child, because each adult covers seven times the time span of a child.

Returning to stars, millisecond pulsars (the "adults") must have a very long life before they turn off; for purposes of this calculation I shall assume that they live for some 20 billion years, which is the age of the universe. Low-mass X-ray binaries (the "children") are estimated to have a relatively short life in comparison, about 100 million years. The statistics now indicate that the ratio of millisecond pulsars to low-mass X-ray binaries should be roughly 20 billion to 100 million, or 200 to 1.

The conclusion is consistent with observation. I mentioned above that a total of three millisecond pulsars have been found so far. More are certain to be discovered, and so it is reasonable to expect that there are probably several thousand millisecond pulsars lurking undetected in the galaxy. (The pulsars cannot be seen from the earth unless they are close enough.) One would therefore expect to see a few dozen low-mass X-ray binaries in the galaxy, which is indeed what one finds.

Statistical considerations are naturally reassuring, but perhaps more important to us is that we were able to employ only basic principles of physics to develop a model that accurately predicts how fast the rate of spin of pulsar 1937 + 214 should change. According to our theory, 1937 + 214 is a resurrected radio pulsar. It began as a binary and was spun up by accretion. When the accretion stopped, the star had a high enough spin to become a radio pulsar in spite of its relatively weak magnetic field. We determined that its period should increase at the rate of 10^{-20} to 10^{-19} second per second, which is many orders of magnitude less than what one might expect for any fast pulsar known before 1982. Our prediction has been confirmed: the measured value of the rate at which the period increases is 1.24×10^{-19} second per second, with an error of $.25 \times 10^{-19}$ second per second.

A complete resolution of the mystery of 1937 + 214 is not yet in hand, however. In particular, if the pulsar was spun up by accretion, where is its companion star? The question became more vexing in 1983 with the discovery of the 6.133-millisecond pulsar 1953 + 29 (by Valentin Boriakoff of Cornell University, R. Buccheri of the National Research Council of Italy in Palermo and F. Fauci of the University of Palermo), which does have a binary companion.

Soon after the discovery, Helfand, Ruderman and I set out to ponder why 1953 + 29 should be part of a binary system and 1937 + 214 should be isolated. We made use of the fact that there are two types of low-mass X-ray binaries: bright ones and weak ones. Bright binaries can be understood only in terms of a "normal" star swelling up to become a red giant and subsequently losing its outer envelope

Figure 12.5 MILLISECOND PULSAR designated 1937 + 214 is believed to be a resurrected, or recycled, dense neutron star. Perhaps some dozens of million years ago or more the star was part of an X-ray binary: a spinning neutron star orbited by a companion star (*top*). As the stars orbited each other, the neutron star slowly accreted matter from its companion. The impact of the incoming matter on the surface of the star released X rays, and the star began to spin faster. When the accretion eventually stopped, and after the companion star had accreted itself "to death," the neutron star was presumably spinning fast enough to be a radio pulsar (*bottom*). Here the red wavy lines represent radio waves; yellow, visible light; green, X rays; and blue, gamma rays. At present it is not understood why the companion star should have disappeared.

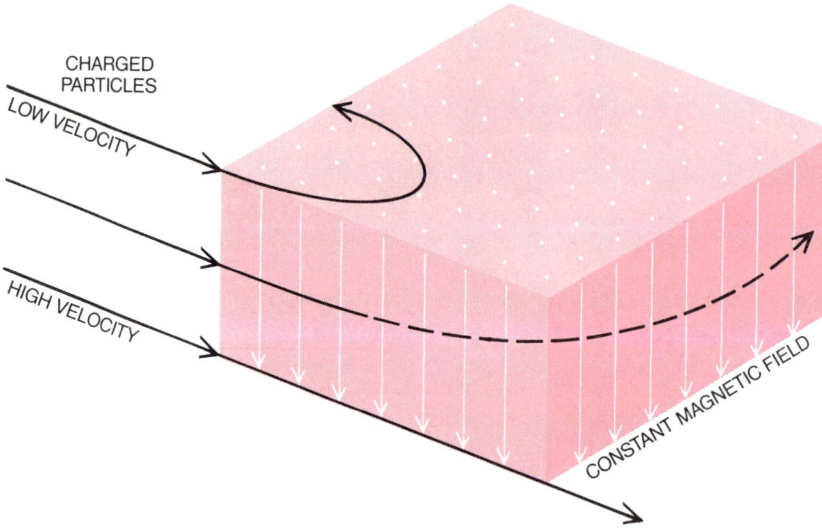

Figure 12.6 CHARGED PARTICLE moving in a direction perpendicular to a magnetic field is deflected by the so-called Lorentz force in such a way that it would "like" to move in a circle. As long as the magnetic field is weak and the particle has high velocity, the deflection is not substantial (*bottom*). As the ratio of field strength to velocity increases, larger deflections occur (*middle*), until finally the particle is turned back by the field (*top*).

of material to a companion neutron star. A weak binary, in contrast, might result if two stars were drawn together by losing orbital energy to gravitational radiation.

In the past five years or so a number of investigators have shown that if a bright low-mass X-ray binary gives birth to anything, it can only be a binary system—in this case a pulsar and a leftover core of the "normal" star. (Among the investigators reaching this conclusion are Paul C. Joss and Saul A. Rappaport of the Massachusetts Institute of Technology, Edward P. J. van den Heuvel and G. J. Savonije of the Astronomical Institute in Amsterdam, Ronald E. Taam of Northwestern University and Ronald F. Webbink of the University of Illinois at Urbana-Champaign.) We consequently hypothesized that 1953 + 29 may have started out as a bright low-mass X-ray binary and 1937 + 214 may have originally been a weak one.

The situation became more complex in 1986 with the discovery of the 5.362-millisecond pulsar 1855 + 09 (by D. J. Segelstein, L. A. Rawley, D. R. Stinebring, A. S. Fruchter and Joseph H. Taylor of Princeton University), which also has a binary companion. The histories of 1855 + 09 and 1953 + 29 appear in principle quite similar both to each other and to those of five other binary pulsars that, while they are not superfast, have periods of several tens of milliseconds. As a result 1937 + 214 really begins to be conspicuous in not having a companion. Several scenarios to account for how the companion might have disappeared have been proposed (including at least one in which the pulsar is held to be relatively young, about a million years old). To date, however, none is entirely successful, and the quest for a satisfactory explanation continues.

Perhaps the most crucial missing link in our story was and still is the fact that no observed low-mass X-ray binary seems as yet to have exhibited a superfast neutron star as one of its two members. Indeed, if a low-mass X-ray binary is to give birth to a superfast pulsar, one would expect it to contain a rotating neutron star that has a period of a few milliseconds. Why have no such periodicities been seen?

Several reasons suggest themselves. The most plausible one pertains to the fact that any neutron star found in a low-mass X-ray binary is likely to be quite old. Its magnetic field will therefore be a relic of an earlier and stronger magnetic field, which is just what one needs to get millisecond periods by

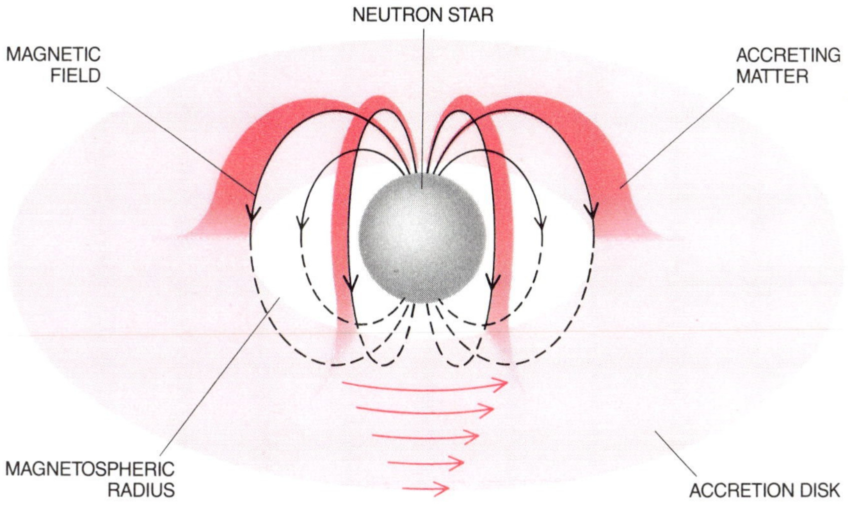

Figure 12.7 ACCRETION of material onto a neutron star occurs primarily on the polar caps because it is there that most magnetic field lines begin and terminate. Charged particles moving parallel to the field lines are not deflected, and so they tend to accumulate at the poles. Charged particles moving perpendicular to the field lines are deflected, however, although collisions with other incoming particles keep them from turning back. The point at which the particles tend to be stopped is called the magnetospheric radius. In the region inside, "the magnetosphere," particles move mostly along field lines and rotate with the star.

the accretion process. David S. Eichler and Zhengzhi Wang of the University of Maryland at College Park have shown that the strength of the relic magnetic field varies widely over the surface. Consequently whereas an ordinary neutron star has two magnetic X-ray hot spots (polar caps from which X rays emanate), an old neutron star may have many X-ray hot spots on its surface.

If the star rotates and an X-ray telescope is aimed at it, the hot spots will show up as variations of X-ray intensity. The variations will not be very strong or sharp, however; they will look like erratic flickerings at much higher frequencies. Even though the intensity pattern will repeat once per stellar revolution, it will be so smeared out that the period, and hence the spinning star itself, may not be detectable against the background of X-ray noise from that source.

Other effects might also make the reputed rotation of the neutron star of a low-mass X-ray binary difficult to see. Accreting matter from the companion star might scatter the X rays from the hot spots, thereby reducing the amount of modulation. In addition the neutron star probably acts as a "gravitational lens": it bends the X rays coming out of it, further reducing the degree of modulation.

Even though such considerations are well taken, one would still be glad to actually find a spinning neutron star in a low-mass X-ray binary. In 1983 the European Space Agency launched *EXOSAT,* an X-ray observatory satellite, into orbit around the earth. One of the experiments on the satellite was to systematically search for very short X-ray periods in low-mass X-ray binaries. Although no single well-defined period was actually found, some quasi periodicities, or intermittent flickerings of radiation, were picked up in the galactic source called GX5-1. The time between "bursts" ranged from 25 to 50 milliseconds.

The quasi periodicities have some unusual properties, as Michel van der Klis of the European Space Agency and his colleagues (F. Jansen of the Laboratory of Space Research, J. van Paradijs and M. Sztajno of the Astronomical Institute in Amsterdam, J. Trümper of the Max Planck Institute for Physics and Astronomy in Munich and W. H. G. Lewin of M.I.T. and the Max Planck Institute) reported in International Astronomical Union circular No. 4043 on March 13, 1985, and subsequently in *Nature.* First, the quasi periodicities are not constant; in fact, they show large variability. The quasi-periodic signal is strongest at the longest quasi period

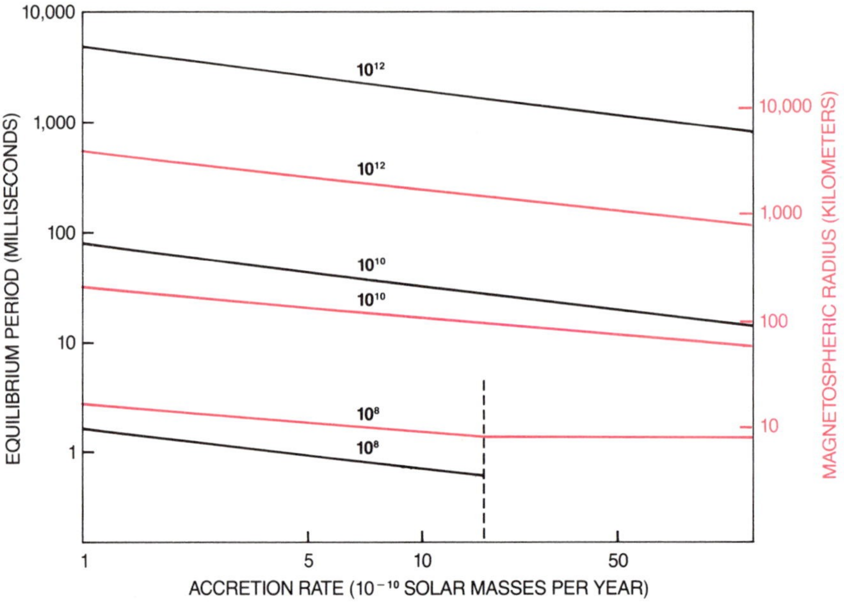

Figure 12.8 EQUILIBRIUM PERIOD AND MAGNETOSPHERIC RADIUS are plotted as a function of accretion rate for various magnetic field strengths. The equilibrium period is the period at which matter at the magnetospheric radius rotates with the star. The calculations are for a 1.4-solar-mass star (a star that has a mass 1.4 times that of the sun).

observed and falls off consistently for shorter periods. The most intriguing feature of the phenomenon is that the total luminosity of GX5-1 correlates directly with the length of the period: the shorter the period is, the brighter the source shines.

Several weeks before the IAU circular came out, Alpar, who was then at Illinois, called to tell me about it. A friend of his in Europe had attended a seminar on the findings, and the rumors then spread. As bizarre as the findings were, they sounded genuine, and we knew they contained a significant clue to our story of the superfast pulsars. We worked hectically to develop a model that would account for the observations.

In our model the quasi periodicities resulted from matter revolving about the neutron star and attempting to "hook" onto the magnetic field, which rotates with the star, of course. The "bursts" in themselves come at a frequency that equals the difference between the orbital frequency of the accreting matter and the rotating frequency of the star. The spread in the rates of orbital frequencies leads to quasi-periodic behavior rather than well-defined behavior. According to our analysis, GX5-1 must spin at a rate of about six to 10 milliseconds and have a surface magnetic field of roughly a billion gauss. We became excited indeed, because both values closely match those of the binary pulsar 1953 + 29.

How wonderful the news was! At last there was a clear indication of a millisecond periodicity inside an active low-mass X-ray source. The entire picture of the formation of millisecond pulsars seemed to come together.

X-ray astronomy has been thrown into a whirlwind with the discovery of the quasi-periodic X-ray source GX5-1, which is now known most generally as a quasi-periodic oscillator. To date more than 10 quasi-periodic oscillators have been discovered, of which seven exhibit short periods (from 25 to 250 milliseconds). All seven are associated with highly luminous low-mass X-ray binaries. The textbook behavior of GX5-1 has turned into a large zoo of phenomena in the other quasi-periodic oscillators. All the complications are certain to trigger further discussion and modeling of quasi-periodic oscillators and millisecond pulsars in coming years.

Much more remains to be done. Yet a fascinating picture of the resurrection of superfast, long-lived pulsars from ancient neutron stars is emerging. The picture brings with it a new perspective on some of the oldest stellar systems in the galaxy. Pulsars have already led to fresh insights in such areas as particle, nuclear, solid-state, plasma and superfluid physics as well as electromagnetism and general relativity. Now the oldest pulsars light up the path of evolution of the galaxy. "Twinkle, twinkle, little star, we still wonder what you are."

The Editor

DONALD E. OSTERBROCK is professor of astronomy and astrophysics at Lick Observatory of the University of California, Santa Cruz, and was its director from 1973 to 1981. After serving in the U. S. Army Air Force in World War II, he studied at the University of Chicago and its Yerkes Observatory, receiving his B.S. in 1948, his M.S. in 1949, and his Ph.D. in astronomy in 1952. He then did postdoctoral work at Princeton University and was a faculty member at the California Institute of Technology and the University of Wisconsin before going to Lick Observatory.

The Authors

MICHAEL HOSKIN ("William Herschel and the Making of Modern Astronomy") is head of the department of history and philosophy of science at the University of Cambridge and president of Churchill College at Cambridge. He studied mathematics at the University of London, getting a B.A. in 1951 and an M.A. in 1952. He received a Ph.D in algebraic geometry from Cambridge in 1956. He was a history of science lecturer at the University of Leicester from 1957 to 1959, when he joined the faculty at Cambridge.

DAVID H. DeVORKIN ("Henry Norris Russell") is curator of the history of astronomy at the National Air and Space Museum of the Smithsonian Institution in Washington, D.C. He holds degrees in astronomy and astrophysics from the University of California, Los Angeles, San Diego State University and Yale University and a 1978 Ph.D. in the history of science from the University of Leicester and is the author of *Race to the Stratosphere: Manned Scientific Ballooning in America,* (Springer-Verlag, 1989).

THOMAS R. GEBALLE ("The Central Parsec of the Galaxy") is on the staff of the Infrared Astronomy Centre in Hawaii. He studied at the University of Washington, University of Amsterdam and University of California, Berkeley, where he received his bachelor's degree in 1967 and his Ph.D. in physics in 1974. Following a postdoctoral year at Berkeley, he spent two years at the University of Leiden, returning to the U.S. to accept a fellowship at the Hale Observatories.

IVAN R. KING ("Globular Clusters") is professor of astronomy at the University of California, Berkeley. He got his A.B. at Hamilton College in 1947 and his Ph.D. from Harvard University in 1952. After working for the Department of Defense he took a position in the astronomy department at the University of Illinois at Urbana-Champaign. He remained there for eight years and then moved to Berkeley in 1964. As a U.S. member of the European Faint Object Camera team, he has been active in the preparations for observations with the Space Telescope.

NICK SCOVILLE and **JUDITH S. YOUNG** ("Molecular Clouds, Star Formation and Galactic Structure") are both astronomers. Scoville is professor of astronomy at the California Institute of Technology. He earned his Ph.D. in astronomy from Columbia University, then did postdoctoral work at the University of Minnesota and at Cal Tech. From 1975 to 1983 he was professor of astronomy at the University of Massachusetts at Amherst. Young is a member of the department of physics and astronomy at the University of Massachusetts. Her B.A. in astronomy is from Harvard University and her Ph.D. in physics is from the University of Minnesota.

KLAAS S. DE BOER and **BLAIR D. SAVAGE** ("The Coronas of Galaxies") are both astrophysicists. De Boer received his education at the University of Groningen and was a member of the faculty from 1971 to 1974. From 1974 to 1977 he was a member of the Netherlands Space Research Group and served part-time on the Astronomical Netherlands Satellite ultraviolet team. In 1981 he joined the faculty of the University of Tübingen. Savage's bachelor's degree in engineering (1964) is from Cornell University. His M.A. (1966) and Ph.D. (1967) are from Princeton University. In 1967 and 1968 he was research associate at Princeton. In 1968 he went to the University of Wisconsin at Madison, where he became professor in 1978.

MARGHERITA HACK ("Epsilon Aurigae") has been director of the Astronomical Observatory of Trieste since 1964. She was granted a Ph.D. in physics from the University of Florence in 1945. In 1947 she joined the Astrophysical Observatory of Arcetri in Florence and in 1954 she moved to the Astronomical Observatory of Brera at Milan-Merate. Hack is also director of the Italian popular magazine *L'Astronomia.*

JOHN S. MATHIS, BLAIR D. SAVAGE and **JOSEPH P. CASSINELLI** ("A Superluminous Object in the Large Cloud of Magellan") are professors of astronomy at the University of Wisconsin at Madison. Mathis got a B.S. in physics at the Massachusetts Institute of Technology and a Ph.D. in astronomy from the California Institute of Technology. He spent two years on the faculty of Michigan State University before joining the Wisconsin faculty. Savage received his B.S. at Cornell University. His M.A. and his Ph.D. in astronomy are from Princeton University. After a year as research associate at Princeton he moved to Wisconsin in 1968. Cassinelli had his undergraduate education at Xavier University and earned his M.S. in physics at the University of Arizona and his Ph.D. in astronomy from the University of Washington. He was research associate at the Joint Institute of Laboratory Astrophysics for two years before joining Wisconsin in 1972.

HANS A. BETHE and **GERALD BROWN** ("How a Supernova Explodes") are, respectively, professor emeritus of physics at Cornell University and professor of physics at the State University of New York at Stony Brook. Bethe was educated in Germany. In 1935 he joined the faculty at Cornell and in 1967 was awarded the Nobel Prize in physics for his descriptions of the nuclear reactions through which stars generate energy. Brown earned his Ph.D. at Yale University in 1950. He taught at the University of Birmingham, Nordic Institute for Theoretical Atomic Physics in Copenhagen and Princeton University before joining the faculty at Stony Brook in 1968.

FREDERICK D. SEWARD, PAUL GORENSTEIN and **WALLACE H. TUCKER** ("Young Supernova Remnants") have worked together for many years in the field of high-energy astronomy. Seward is an astrophysicist at the Smithsonian Astrophysical Observatory in Cambridge, Mass., where he directs the Einstein Observatory Data Bank. He holds a bachelor's degree from Princeton University and a doctorate in nuclear physics from the University of Rochester (1958). Gorenstein is also an astrophysicist at the Smithsonian Astrophysical Observatory. He is a graduate of Cornell University and the Massachusetts Institute of Technology, which granted him a Ph.D. in physics (1962). Tucker divides his time between the Smithsonian Astrophysical Observatory, where he is a senior theoretician, and the University of California, Irvine, where he is a visiting lecturer. He was educated at the University of Oklahoma and the University of California, San Diego, where he received a Ph.D. in physics (1966).

STAN WOOSLEY and **TOM WEAVER** ("The Great Supernova of 1987) have collaborated for more than 15 years. Woosley is professor of astronomy and astrophysics at the University of California, Santa Cruz and chair of the department; he is also a member of the General Studies Group in the physics department of the Lawrence Livermore National Laboratory. He received his B.A. (1966), M.Sc. (1969) and Ph.D. (1971) from Rice University at Houston. Weaver heads the General Studies Group at Livermore and is chief scientist of the X-ray laser program. He received his B.S. from the California Institute of Technology (1971) and his M.A. (1972) and Ph.D. (1975) from the University of California, Berkeley.

JACOB SHAHAM ("The Oldest Pulsars in the Universe") is a professor of physics at Columbia University. He got his B.Sc. (1963), M.Sc. (1965) and Ph.D. (1971) from the Hebrew University of Jerusalem. After three years in a postdoctoral assignment at the University of Illinois at Urbana-Champaign, he returned to Jerusalem and worked for 12 years at the Racah Institute of Physics.

Bibliographies

1. William Herschel and the Making of Modern Astronomy

Bennett, J. A. 1976. On the power of penetrating into space: The telescopes of William Herschel. *Journal for the History of Astronomy* 7 (June): 75–108

———. 1982. Herschel's scientific apprenticeship and the discovery of Uranus. In *Uranus and the outer planets,* ed. Garry E. Hunt. Cambridge University Press.

Porter, Roy. 1982. William Herschel, Bath, and the Philosophical Society. In *Uranus and the outer planets,* ed. Garry E. Hunt. Cambridge University Press.

2. Henry Norris Russell

Menzel, Donald H. 1972. The history of astronomical spectroscopy. *Annals of the New York Academy of Sciences* 198:225–244.

Hufbauer, Karl. 1981. Astronomers take up the stellar energy problem. *Historical Studies in the Physical Sciences* 11 (March): 277–303.

DeVorkin, David H., and Ralph Kenat. 1983. Quantum physics and the stars. *Journal for the History of Astronomy* 14 (June): 102:132 and 14 (October): 180–222.

DeVorkin, David H. 1984. Stellar evolution and the origins of the Hertzsprung-Russell diagram in early astrophysics. In *Astrophysics and twentieth-century astronomy to 1950,* ed. Owen Gingerich. Cambridge University Press.

3. The Central Parsec of the Galaxy

Oort, J. H. 1977. The galactic center. *Annual Review of Astronomy and Astrophysics* 15:295–362.

Becklin, E. E., K. Matthews, G. Neugebauer and S. P. Willner. 1978. Infrared observations of the galactic center. *The Astrophysical Journal* 219 (January 1): 121–128.

Rieke, G. H., C. M. Telesco and D. A. Harper. 1978. The infrared emission of the galactic center. *The Astrophysical Journal* 220 (March 1): 556–567.

4. Globular Clusters

Hanes, D., and B. Madore, beds. 1980. *Globular clusters.* Cambridge University Press.

Hesser, James E., ed. 1980. *Star clusters.* D. Reidel Publishing Company.

King, Ivan R. 1981. The dynamics of globular clusters. *The Quarterly Journal of the Royal Astronomical Society* 22:227–243.

5. Molecular Clouds, Star Formation and Galactic Structure

Wynn-Williams, Gareth. 1981. The newest stars in Orion. *Scientific American* 245 (August): 46–55.

Morris, Mark, and L. J. Rickard. 1982. Molecular clouds in galaxies. *Annual Review of Astronomy and Astrophysics* 20:517–545.

Wynn-Williams, C. G. 1982. The search for infrared protostars. *Annual Review of Astronomy and Astrophysics* 20:587–618.

Blitz, Leo. 1982. Giant molecular-cloud complexes in the galaxy. *Scientific American* 246 (April): 84–94.

Scoville, Nick, and Judith S. Young. 1983. The molecular gas distribution in M51. *The Astrophysical Journal* 265 (February 1): 148–165.

6. The Coronas of Galaxies

Spitzer, Lyman, Jr. 1956. On a possible interstellar galactic corona. *The Astrophysical Journal* 124 (July): 20–34.

Savage, Blair D., and Klaas S. de Boer. 1981. Stellar gas at large distances from the galactic plane. *The Astrophysical Journal* 243 (January 15): 460–484.

Spitzer, Lyman, Jr. 1982. *Searching between the stars.* Yale University Press.

York, D. G. 1982. Gas in the galactic halo. *Annual Review of Astronomy and Astrophysics.* 20: 221–248.

7. Epsilon Aurigae

Kuiper, G. P., O. Struve and B. Strömgren. 1937. The interpretation of ϵ Aurigae. *The Astrophysical Journal* 86 (December): 570–612.

Kraft, Robert P. 1954. The atmosphere of the I component of Epsilon Aurigae. *The Astrophysical Journal* 120 (November): 391–400.

Huang, Su-Shu. 1965. An interpretation of Epsilon Aurigae. *The Astrophysical Journal* 141 (April 1): 976–984.

Wright, K. O. 1970. The Zeta Aurigae stars. *Vistas in Astronomy* 12:147–182.

Sahade, Jorge, and Frank Bradshaw Wood. 1978. *Interacting binary stars.* Pergamon Press.

Reddy, Francis J. 1982. The mystery of Epsilon Aurigae. *Sky and Telescope* 63 (May): 460–462.

8. A Superluminous Object in the Large Cloud of Magellan

Cassinelli, Joseph P., John S. Mathis and Blair D. Savage. 1981. Central object of the 30 Doradus nebula, a supermassive star. *Science* 212 (June 26): 1497–1501.

Savage, Blair D., Edward L. Fitzpatrick, Joseph P. Cassinelli and Dennis C. Ebbets. 1983. The nature of R136A, the superluminous central object of the 30 Doradus nebula. *The Astrophysical Journal* 273 (October 15): 597–623.

Humphreys, Roberta M., and Kris Davidson. 1984. The most luminous stars. *Science* 223 (January 20): 243–249.

9. How a Supernova Explodes

Burbidge, E. Margaret, William A. Fowler and F. Hoyle. 1957. Synthesis of the elements in stars. *Reviews of Modern Physics* 29 (October): 547–650.

Rees, M. J., and R. J. Stoneham. 1982. *Supernovae: A survey of current research.* D. Reidel Publishing Co.

Brown, G. E., H. A. Bethe and Gordon Baym. 1982. Supernova theory. *Nuclear Physics* A375 (February 15): 481–532.

Trimble, Virginia. 1982. Supernovae, part I: The events. *Reviews of Modern Physics* 54 (October): 1183–1224.

———. 1983. Supernovae, part II: The aftermath. *Reviews of Modern Physics* 55 (April): 511–563.

10. Young Supernova Remnants

Dreyer, J. L. E. 1890. *Tycho Brahe, a picture of scientific life and work in the sixteenth century.* A. & C. Black.

Clark, David H., and F. Richard Stephenson. 1977. *The historical supernovae.* Pergamon Press.

Danziger, John, and Paul Gorenstein, eds. 1983. *Supernova remnants and their x-ray emission.* D. Reidel Publishing Co.

11. The Great Supernova of 1987

Woosley, S. E., and Thomas A. Weaver. 1986. The physics of supernova explosions. *Annual Review of Astronomy and Astrophysics* 24:205–253.

Woosley, S. E., and M. M. Phillips. 1988. Supernova 1987A! *Science* 240 (May 6): 750–759.

Kafatos, M., and A. Michalitsianos. 1988. *Supernova 1987A in the Large Magellanic Cloud.* Cambridge University Press.

Marschall, Laurence A. 1988. *The supernova story.* Plenum Press.

Arnett, W. David, John N. Bahcall, Robert P. Kirshner and Stanford E. Woosley. 1989. Supernova 1987A. *Annual Review of Astronomy and Astrophysics* 27:629–700.

12. The Oldest Pulsars in the Universe

Backer, D. C., Shrinivas R. Kulkarni, Carl Heiles, M. M. Davis and W. M. Goss. 1982. A millisecond pulsar. *Nature* (December 16): 615–618.

Alpar, M. A., A. F. Cheng, M. A. Ruderman and J. Shaham. 1982. A new class of radio pulsars. *Nature* 300 (December 16): 728–730.

Shapiro, Stuart L., and Saul A. Teukolsky. 1983. *Black holes, white dwarfs and neutron stars: The physics of compact objects.* John Wiley & Sons, Inc.

Van der Klis, M., F. Jansen, J. van Paradijs, W. H. G. Lewin, E. P. J. van den Heuvel, J. E. Trümper and M. Sztajno. 1985. *Nature* 316 (July 18): 225–230.

Sources of the Photographs

Introduction. Lick Observatory: Figures I.1 and I.2

Sky Publishing Corporation: Figure 1.1

Quesada/Burke, courtesy of the New York Public Library: Figures 1.2, 1.4 (*top*) and 1.6

The Granger Collection: Figure 1.3

Royal Astronomical Society: Figures 1.4 (*bottom*) and 1.5

W. F. Meggers Collection, Niels Bohr Library, American Institute of Physics: Figure 2.1

Andrew Christie: Figure 2.3 (*right*)

Owen Gingerich: Figure 2.4

Scientific American, Inc.: Figure 2.5

Hale Observatories: Figure 3.1

E. S. Light, R. E. Danielson and Martin Schwarzschild of Princeton University, courtesy of Hale Observatories. Figure 3.2

Copyright © 1983 California Institute of Technology: Figure 4.1 (*top*) Ivan R. King, University of California, Berkeley: Figures 4.1 (*bottom*) and 4.5

Stanislav Djorgovski, University of California, Berkeley: Figure 4.6

Kitt Peak National Observatory: Figures 5.1 and 5.2 (*top left*)

Nick Scoville: Figures 5.2 (*top right*) and 5.3

Lick Observatory: Figure 5.2 (*bottom left*)

Gareth Wynn-Williams: Figure 5.2 (*bottom right*)

Nick Scoville and Judith S. Young: Figures 5.6 (*left*) and 5.8 (*bottom*)

James Smith: Figure 5.6 (*right*)

Hale Observatories: Figure 5.8 (*top*)

Kwok-Yung Lo: Figure 5.9

Renzo Sancisi, Kapteyn Laboratories: Figure 6.2

Klaas S. de Boer and Blair D. Savage: Figure 6.4

Cerro Tololo Inter-American Observatory: Figure 8.1

John S. Mathis, Blair D. Savage and Joseph P. Cassinelli, University of Wisconsin at Madison: Figures 8.2, 8.3 and 8.8

Mendillo Collection of Astronomical Prints: Figure 10.1

Frederick D. Seward, Paul Gorenstein and Wallace H. Tucker: Figure 10.2

David Green and Steven Gull: Figure 10.5

David F. Malin, Anglo-Australian Observatory: Figure 11.1

John Maduell, Lawrence Livermore National Observatory: Figure 11.3

Arthur Francheschi, Lawrence Livermore National Observatory: Figure 11.6

INDEX

Page numbers in *italics* indicate illustrations.